T0134967

Springer Theses

Recognizing Outstanding Ph.D. Research

Aims and Scope

The series "Springer Theses" brings together a selection of the very best Ph.D. theses from around the world and across the physical sciences. Nominated and endorsed by two recognized specialists, each published volume has been selected for its scientific excellence and the high impact of its contents for the pertinent field of research. For greater accessibility to non-specialists, the published versions include an extended introduction, as well as a foreword by the student's supervisor explaining the special relevance of the work for the field. As a whole, the series will provide a valuable resource both for newcomers to the research fields described, and for other scientists seeking detailed background information on special questions. Finally, it provides an accredited documentation of the valuable contributions made by today's younger generation of scientists.

Theses are accepted into the series by invited nomination only and must fulfill all of the following criteria

- They must be written in good English.
- The topic should fall within the confines of Chemistry, Physics, Earth Sciences, Engineering and related interdisciplinary fields such as Materials, Nanoscience, Chemical Engineering, Complex Systems and Biophysics.
- The work reported in the thesis must represent a significant scientific advance.
- If the thesis includes previously published material, permission to reproduce this must be gained from the respective copyright holder.
- They must have been examined and passed during the 12 months prior to nomination.
- Each thesis should include a foreword by the supervisor outlining the significance of its content.
- The theses should have a clearly defined structure including an introduction accessible to scientists not expert in that particular field.

More information about this series at http://www.springer.com/series/8790

Reza Hassanli

Behavior of Unbounded Post-tensioned Masonry Walls

Doctoral Thesis accepted by
University of South Australia, Adelaide, Australia

 Springer

Author
Dr. Reza Hassanli
School of Natural and Built Environment
University of South Australia
Adelaide, SA, Australia

Supervisors
Dr. Mohamed ElGawady
Missouri University of Science and
   Technology
Rolla, MO, USA

Prof. Julie Mills
School of Natural and Built Environment
University of South Australia
Adelaide, SA, Australia

ISSN 2190-5053   ISSN 2190-5061   (electronic)
Springer Theses
ISBN 978-3-030-06723-6   ISBN 978-3-319-93788-5   (eBook)
https://doi.org/10.1007/978-3-319-93788-5

This Springer imprint is published by the registered company Springer Nature Switzerland AG
The registered company address is: Gewerbestrasse 11, 6330 Cham, Switzerland

This thesis is dedicated to my wonderful parents and wife

For their endless love, support and encouragements

Supervisor's Foreword

The thesis entitled "Behavior of Unbonded Post-Tensioned Masonry Walls" covers an important mechanism in a new generation of structural walls, called "Self-centering". A Self-centering response can be introduced to structures by using unbonded post-tensioning (PT) steel. In self-centering systems, the restoring nature of the force in the PT steel returns the system back to its original position. This behavior reduces residual drifts and structural damage during earthquake ground motion, and is particularly favorable for structures which are designed for immediate occupancy performance levels. Due to its unique behaviour, the self-centering concept has been applied to various types of structures including masonry walls, however the focus has been more on their out-of-plane behaviour.

Due to limited knowledge about the in-plane behavior of post-tensioned masonry walls (PT-MWs), design codes use a very conservative approach where the elongation of the PT steel is ignored in the evaluation of the flexural capacity. However, this approach results in considerable underestimation of the wall strength. To better understand the behavior of PT-MWs and to develop new expressions to predict the strength of PT-MWs, this study was conducted according to the following steps:

- The test results of PT-MWs available in the literature were collected and the ability of the current code expressions in predicting the strength of the walls was evaluated. Moreover, the seismic response factors including ductility, response modification factor and displacement amplification were determined for these specimens.
- Finite element (FE) models of masonry prisms were developed and calibrated with experimental results, and a parametric study was performed to investigate the accuracy of the height-to-thickness ratio correction factors presented in masonry codes. As a result, a series of strength correction factors was recommended to be considered for concrete masonry prisms based on the size of the prism.

- Using well-validated finite element models, a parametric study was performed to investigate the effect of different parameters on the wall rotation and compression zone length. Multivariate regression analysis was performed to develop a "refined method" to estimate the flexural strength of unbonded PT-MWs, considering the elongation of PT bars.
- Using a validated analytical procedure, a parametric study was performed to obtain the force-displacement response of PT-MWs. Multivariate regression analysis was performed to develop a "simplified method" to predict the flexural strength of unbonded PT-MWs.
- An experimental study including in-plane testing of four PT-MWs was performed with emphasis on: (1) The influence of different parameters on the wall demands, (2) Providing experimental evidence to support the developed design methodology.
- Using both the experimental and FE modelling results, it was shown that both the proposed simplified and refined methods could accurately predict the flexural strength prediction of PT-MWs and can be considered for the design of such walls.

The thesis provides an important advance in the design of a new generation of masonry walls that have significant potential for use in seismic zones.

Adelaide, Australia   Prof. Julie Mills
May 2017

Parts of this thesis have been published in the following journal articles:

- Hassanli R., ElGawady M.A. and Mills J.E., *Simplified approach to predict the flexural strength of unbonded post-tensioned masonry walls,* Journal of Engineering Structures, Volume 142, pp 255–271, 2017.
- Hassanli R., ElGawady M.A. and Mills J.E., *In-plane flexural strength of unbonded post-tensioned concrete masonry walls,* Journal of Engineering Structures, 136, 245–260, 2017.
- Hassanli R., ElGawady M.A. and Mills J.E., *Experimental investigation of cyclic in-plane behavior of unbonded post-tensioned masonry walls',* Journal of Structural Engineering, 142(5), 2016.
- Hassanli R., ElGawady M.A. and Mills J.E., *Strength and seismic performance factors of post-tensioned masonry walls,* Journal of Structural Engineering, 141 (11), 2015.
- Hassanli R., ElGawady M.A. and Mills J.E., *Effect of dimensions on the compressive strength of concrete masonry prisms,* Advances in Civil Engineering Materials, ASTM, 4(1), 175–201, 2015.

Acknowledgements

The author would like to express his sincere gratitude and appreciation to his supervisors Dr. Mohamed ElGawady and Prof. Julie Mills for their continuous guidance, support and inspiration throughout every single step in this research work.

The authors gratefully acknowledge the support of Bianco precast company, Boral Company and the University of South Australia in experimental work. Special thanks to Mr. T. Benn, Mr. T. Golding, Mr. W. Penney and Dr. H. Senko and the technical support staff from the Concrete Laboratory at the University of South Australia, for their dedication and assistance before and during the experimental study.

The author is obliged to the School of Natural and Built Environments at the University of South Australia, the Australian Building Code Board (ABCB), Bianco precast and Boral Company, for the financial support provided to this research work.

The author wishes to express his sincere gratitude to his family, especially his parents and wife for their continuous support and encouragement.

Contents

Chapter 1
Introduction

Masonry is probably the oldest and one of the most widely used man-made construction materials in the world (Ganz 1990; Bean Popehn and Schultz 2003; Drysdale and Hamid 2005). The major advantage of masonry is that in terms of raw materials it is highly available worldwide and in terms of construction it is easy and economical. Moreover, masonry is a highly durable, fire resistant and sound absorbing material.

In recent decades, the use of unreinforced masonry has declined in comparison with other materials such as concrete and steel, mainly due to the limited application of masonry as a structural material and also the complex behavior of masonry.

Improved materials, refined innovative design procedures and innovative construction ideas, together with the inherent advantages of masonry such as availability and easy construction, help to make masonry a unique and useful construction material.

1.1 Reinforced Masonry

Masonry has relatively high compressive strength, so it has mainly been used in construction to carry gravitational loads due to the permanent and imposed loads on a structure, which result in compression stress in the masonry. However, the main drawback of masonry is that it is weak in tension. Conventional unreinforced masonry walls, exhibit poor behavior under in-plane and out-of-plane loads and moments, usually due to the crack propagation at the mortar-unit interface. This results in a brittle type of failure and provides low ductility. The traditional method to overcome the tensile weakness of masonry was to rely on the mass of the wall. To prevent overturning and keep the compressive resultant force within the section to avoid eccentricity, ancient walls were usually wider at the base. However, this method is not economical and not appealing in terms of seismic design for modern construction.

© Springer International Publishing AG, part of Springer Nature 2019
R. Hassanli, *Behavior of Unbounded Post-tensioned Masonry Walls*,
Springer Theses, https://doi.org/10.1007/978-3-319-93788-5_1

To overcome the disadvantage of masonry in tension and to make a cost-effective construction material, researchers, engineers and designers started to incorporate steel in masonry structures. In reinforced masonry the steel carries the flexural tensile stresses and also contributes to the shear strength while the masonry withstands the compressive stresses, the same concept considered in reinforced concrete design. In reinforced masonry the unit cavities along the steel bars must be grouted, to ensure the wall maintains its integrity and the steel deforms with the wall. Reinforced masonry walls can be fully or partially grouted (Fig. 1.1). From a construction viewpoint, partially grouted masonry (PGM) walls are more efficient

Fig. 1.1 Masonry shear wall **a** unreinforced **b** partially grouted, and **c** fully grouted

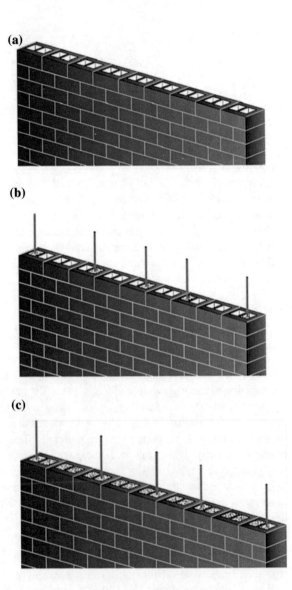

(a)

(b)

(c)

than fully grouted walls, since service installation and construction are easier and faster. Reduction in weight of the structure and saving material are other advantages of PGM walls over fully grouted walls (Minaie 2009).

1.2 Post-tensioned Masonry

The main disadvantage of reinforced masonry is that the steel does not contribute to the strength before tensile cracks occur in the wall. To overcome this drawback, the steel can be post-tensioned. In a post-tensioned wall, a pre-compression force is applied to the wall using post-tensioned steel bars/strands which are placed within the cavities of the masonry. Post-tensioning offers architects and engineers the possibility of actively introducing any desired axial load to the masonry wall to enhance their strength, cracking behavior and ductility (Ganz 2003). The post-tensioning effectively connects the masonry units and provides structural integrity of the wall system. Post-tensioning of masonry is most beneficial in situations in which the lateral load, which can be due to wind load, seismic load or earth/water pressure, is comparatively high and the axial load is low (Bean Popehn et al. 2007). A post-tensioned masonry wall (PT-MW) competes economically with a reinforced concrete wall when the wall's height is greater than three meters, mainly because of formwork savings (Ganz 2003).

Post-tensioning is not only effective for new masonry structures, but it is also a viable method of rehabilitation of masonry structures. It is a particularly favorable approach for strengthening historical structures and monuments. The method maintains the architectural aspects, aesthetic features and historic values while strengthening the structure. Various technical solutions have been introduced to implement seismic retrofitting of unreinforced masonry structures. As an example Fig. 1.2 presents a view of the General Post Office (GPO) building in Sydney, a more than one hundred year old sandstone masonry building. During the

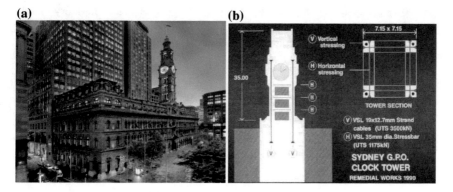

Fig. 1.2 Strengthening of a masonry structure **a** GPO Tower, Sydney General Post Office Building, and **b** Tendon layout in tower (Ganz 1990)

rehabilitation process, this structure underwent a massive restoration both inside and out. As part of the restoration, the GPO Tower shown in Fig. 1.2, was strengthened with four vertical post-tensioning tendons, each included 19×12.7 mm diameter strands; and a number of horizontal pre-stressing bars with a diameter of 35 mm at floor levels (Ganz 2003).

1.3   Unbonded Post-tensioned Masonry Walls

A post-tensioned masonry wall (PT-MW) can be bonded or unbonded. In bonded walls, the ducts through which the PT bars or tendons are passed are grouted using fine aggregate concrete. In unbonded members, the PT ducts are left ungrouted and the PT bars/strands can move freely in the ducts. Recent research has demonstrated that unbonded post-tensioned structural elements including concrete walls, concrete columns, and masonry walls can display high ductility levels while withstanding high levels of seismic loads (Wight et al. 2006; Ryu et al. 2013). When a slender unbonded masonry wall (PT-MW) is subjected to a lateral in-plane load, usually a single horizontal crack forms at the wall-foundation interface. However, in squat unbounded PT-MWs, the failure can be characterized by inclined cracks (shear or flexural-shear cracks) or vertical cracks (due to high compressive stresses in the toe) instead. In a PT-MW with a rocking response, the restoring nature of the post-tensioning (PT) force returns the wall to its original vertical position and minimizes the residual displacement (Fig. 1.3). This behavior is specifically favorable for structures which are designed for immediate occupancy performance levels. The rocking mechanism of PT-MWs results in plastic deformation concentrated at the toe of the wall which can be repaired with minimal cost (Wight 2006; Bean Popehn et al. 2007; ElGawady and Sha'lan 2011; Dawood et al. 2011; Ryu et al. 2013).

Using post-tensioning in masonry walls can make them a competitive construction system. However, to develop design criteria, extensive research must be done to quantify the properties and behavior of post-tensioned masonry (Drake 2004). In particular, more research is required to develop in-depth understanding of the rocking response of unbonded PT walls.

1.4   Aim of the Research

The main aim of this research is to better understand the behavior of unbonded PT-MWs and to develop design guidelines for unbonded PT-MW systems.

Fig. 1.3 Rocking mechanism
in unbonded PT-MW

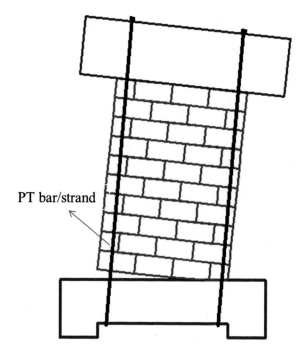

PT bar/strand

1.5 Research Objectives

The objectives of this research are:

(1) To investigate the accuracy of current expressions in predicting the in-plane flexural strength of unbonded PT-MWs, based on experimental and finite element model results.
(2) To provide finite element models to investigate the behavior of masonry prisms.
(3) To develop an analytical procedure to obtain the lateral force-displacement response of unbonded PT-MWs.
(4) To study the relationship between the wall rotation and the compression zone length, and other wall design parameters.
(5) To provide design guidelines and develop expressions to predict the in-plane flexural strength of unbonded PT-MWs.
(6) To investigate the influence of the PT bar spacing and the level of PT axial stress on the wall performance.
(7) To provide experimental evidence to evaluate the proposed design methodology and guidelines.
(8) To develop design criteria for unbonded PT-MWs to ensure ductility and self-centering response.

1.6   Research Questions

The research questions addressed by this research are:

1. What are the parameters influencing the flexural behavior of the unbonded PT-MWs?
2. Is the masonry prism testing methods provided in the current design codes able to accurately reflect the actual strength of masonry?
3. Are the current design codes able to accurately address the ductility, modification factor and displacement amplification factors of PT-MWs?
4. How accurate are the current design expressions in predicting the flexural strength of PT-MWs?
5. What are the limitations of the existing design expressions?
6. Is there an analytical procedure capable of predicting the flexural strength of PT-MWs?
7. How accurate are the proposed design procedure and expressions?
8. How do the axial stress ratio and PT bar spacing affect the seismic behavior of PT-MWs?

1.7   Significance of the Research

Post-tensioning offers a new possibility to innovative engineers and architects for the revival of masonry as a structural material.

While the use of masonry has declined compared with a few decades ago, mainly because of the emergence of concrete and steel, recently, pre-stressing techniques have revitalized masonry as a construction material and form. The main advantage of masonry blocks and bricks is that they are widely available worldwide and the construction is simple. Due to the pre-compression that results from post-tensioning, the inherent deficit of masonry, which is its low tensile strength and cracking under small tensile loads, can be effectively countered.

Extensive applications of pre-stressed masonry were started in the second half of the 20th century (Devalapura et al. 1997a, b). Pre-stressed concrete began to see extensive use in Europe in the 1950s and in North America in the 1960s. Due to the growth of precast concrete and the tendency of fast construction to reduce the construction costs, the technology of pre-stressing has become more appealing recently. There are fundamental similarities in behavior and material properties between concrete and masonry. Concrete blocks can be regarded as precast concrete members and the large potential of using post-tensioning in concrete structures can be extended to masonry structures. As the concept of pre-stressing of masonry, especially unbonded PT masonry, is relatively new, its behavior is not yet well understood. Due to the lack of enough experimental results, more research is required to investigate the seismic behavior of unbonded PT-MWs. The current strength expressions developed for PT-MWs to predict the flexural strength either incorporate simplification, as with the approach of MSJC (2013), where PT forces are considered constant during dynamic excitation, or they are based on semi-empirical equations (Ryu et al. 2013; Wight 2006). The accuracy of these expressions needs to be investigated based on experimental and finite element results. Moreover, the ductility and self-centering behavior of PT-MWs have not been well documented and more experimental research is required to provide a better insight into the behavior of PT-MWs.

(As the MSJC (2013) ignores the elongation of PT bars in predicting the flexural strength of PT-MWs, the method in which the elongation of PT bar is ignored is referred as MSJC (2013) approach in this thesis.)

1.8 Thesis Organization

This thesis has been written using the thesis with publication guidelines and the research papers are utilized as parts of the thesis as shown in Fig. 1.4.

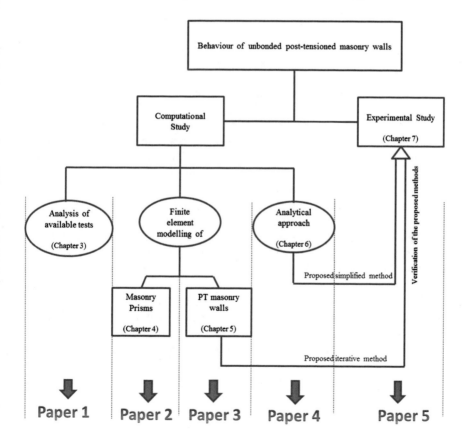

Fig. 1.4 The structure of the thesis organized by publication

The thesis is organized as follows:

Chapter 1: Provides the basic outline of the Ph.D. project by elaborating its significance, research problems, objectives and potential benefits.

Chapter 2: Provides a brief review of research on unbonded post-tensioned masonry walls. The chapter is designed to familiarize readers with the extent of the research that has been conducted to date on PT-MWs.

Chapter 3: The behavior of PT-MWs is investigated using a database of tested walls. The accuracy of ignoring elongation of PT bars which is considered in the current masonry standard joint committee (MSJC 2013) code in evaluating the strength of PT-MWs is studied using the available test results. Using the experimental results, the structural response parameters including ductility, response modification factor and displacement amplification factor are determined for different types of walls including fully grouted, partially grouted, ungrouted walls, walls with confinement plates, walls with supplemental mild steel and walls with an opening.

Chapter 4: Investigates the accuracy of the height-to-thickness ratio (h/t) correction factors presented in the ASTM standard (ASTM C1314-03) and in other international standards using numerical finite element analysis. The FEM is calibrated with experimental results, and then a parametric study is performed to examine the effect of size on the strength of masonry prisms. Calibration of masonry material provided in this chapter is then used in developing finite element models of PT-MWs presented in Chap. 5.

Chapter 5: Develops a design equation to predict the in-plane flexural strength of unbonded PT-MWs. Using well-validated finite element models, a parametric study is performed to investigate the effect of different parameters on the wall rotation and compression zone length, including axial stress ratio, length and height of the wall, initial to yield stress ratio of PT bars and spacing between PT bars. Multivariate regression analysis is performed to develop an equation to estimate the rotation of the unbonded post-tensioned walls at peak strength. Using the drift capacity of the walls and the proposed equation, a design expression and the relevant step-by-step design method is developed to estimate the flexural strength of unbonded PT-MWs, considering the elongation of PT bars. The proposed design expression is also compared with the predicted values obtained considering no elongation of PT bars which is allowed by the MSJC (2013) standard and validated against experimental results as well as finite element model results.

Chapter 6: Develops a simplified design approach to predict the flexural strength of unbonded PT-MWs. The accuracy of different flexural expressions is also investigated in this chapter according to experimental and finite element modelling results. An analytical procedure is developed to predict the force displacement response of PT-MWs. The accuracy of the analytical model is then validated against available experimental test results for unconfined and confined PT-MWs. Using a similar analytical procedure, a parametric study is performed to obtain the force-displacement response of walls with different features. Multivariate regression analysis is performed to develop an empirical equation to estimate the compression zone length in unbonded PT-MWs. The proposed equation for compression zone length is then incorporated into the flexural analysis of post-tensioned masonry walls and validated against experimental results and finite element results.

Chapter 7: Describes the experimental program conducted as a part of this thesis that investigated the behavior of PT-MWs. The accuracy of the MSJC (2013) in ignoring the elongation of PT bars is investigated using the two design equations proposed in Chaps. 5 and 6 to predict the flexural strength of the tested walls. The accuracy of the analytical approach developed in Chap. 6 is verified against the presented experimental results.

Chapter 8: Summarizes the outcomes of this study and proposes a number of recommendations for future research related to this subject.

References

ASTM C1314 (2003) Standard test method for compressive strength of masonry prisms. American Society for Testing and Materials, Pennsylvania, United States

Bean Popehn JR, Schultz AE (2003) Flexural capacity of post-tensioned masonry walls: code review and recommended procedure. PTI J 1(1):28–44

Bean Popehn JR, Schultz AE, Drake CR (2007) Behavior of slender, post-tensioned masonry walls under transverse loading. J Struct Eng 133(11):1541–1550

Dawood H, ElGawady MA, Hewes J (2011) Behavior of segmental precast post-tensioned bridge piers under lateral loads. J Bridge Eng 17(5):735–746

Devalapura RK, Krause GL et al (1997a) Construction productivity advancement research (CPAR) program. Development of an innovative post-tensioning system for prestressed clay brick masonry walls, DTIC document

Devalapura RK, Krause GL, Sweeney SC, Littler D, Staab E (1997b) Construction productivity advancement research (CPAR) program. Development of an innovative post-tensioning system for prestressed clay brick masonry walls. DTIC document

Drake CR (2004) Out-of-plane behavior of slender post-tensioned masonry walls constructed using restrained tendons. M.S. thesis, University of Minnesota. Minnesota, MN, US

Drysdale RG, Hamid AA (2005) Masonry structures: behavior and design. Canadian Masonry and Design Center, Mississauga, Ontario

ElGawady MA, Sha'lan A (2011) Seismic behavior of self-centering precast segmental bridge bents. J Bridge Eng ASCE 16(3):328–339

Ganz HR (1990) Post-tensioned masonry structures. SL Report Series 2, VSL International Ltd., Berne, Switzerland, 35 pp

Ganz HR (2003) Post-tensioned masonry around the world. Concr Int Detroit 25(1):65–70

Masonry Standards Joint Committee (MSJC) (2013) Building code requirements for masonry structures, ACI 530/ASCE 5, TMS 402, American Concrete Institute, Detroit

Minaie E (2009) Behavior and vulnerability of reinforced masonry shear walls. Ph.D thesis, Drexel University, Philadelphia, PA, US

Ryu D, Wijeyewickrema A, ElGawady M, Madurapperuma M (2013) Effects of tendon spacing on in-plane behavior of post-tensioned masonry walls. J Struct Eng 140(4). CID: 04013096

Wight GD (2006) Seismic performance of a post-tensioned concrete masonry wall system. Ph.D. dissertation, Department of Civil and Environmental Engineering, University of Auckland, Auckland, New Zealand

Wight GD, Ingham JM, Kowalsky MJ (2006) Shaketable testing of rectangular post-tensioned concrete masonry walls. ACI Struct J 103(4)

Chapter 2
Literature Review

This chapter provides a brief review of research on unbonded post-tensioned masonry walls. The chapter is designed to familiarize readers with the extent of the research that has been conducted to date on PT-MWs.

By applying some modifications to the pre-stressed concept in concrete to adjust it for masonry, pre-stressed masonry has been used in Australia to enhance the flexural capacity and upgrade existing walls (Page 2001). Although pre-stressed masonry has been used by engineers for many years in Australia, it has had limited applications. In Australia, typically the post-tensioned bars or strands are placed vertically in the masonry cavities. The steel is then post-tensioned and the core grouted, or masonry may be post-tensioned by means of PT plates and anchorage systems (Page 2001).

Mainly due to limited knowledge about pre-stressed masonry behavior, uncertainty about the construction form and liability risk, it has not been extensively used by designers (Laursen 2002).

2.1 Codification of Pre-stressed Masonry

The British masonry standard is the first standard which released the design provisions on post-tensioned masonry (Schultz and Scolforo 1991; Ganz 2003). As the series of research on pre-stressed masonry in 1970s, the writing of draft code provisions for pre-stressed masonry began in the late 1970s and was included in the general masonry code, BS 5628, in 1985 (Wight 2006). The design provisions of pre-stressed masonry were included in the Australian masonry standard, AS3700 (2011), in 1998, based predominantly on the British standard provisions. The code mainly targets construction in non-seismic zones, but can be applied for design in earthquake regions (Laursen 2002).

In the USA, the provisions of pre-stressed prepared masonry by the Masonry Standards Joint Committee in 1999 (MSJC 2005). The Masonry Standards Joint

© Springer International Publishing AG, part of Springer Nature 2019
R. Hassanli, *Behavior of Unbounded Post-tensioned Masonry Walls*,
Springer Theses, https://doi.org/10.1007/978-3-319-93788-5_2

Committee (MSJC) is responsible for the development of the masonry design code. Both AS 3700 and MSJC have typically followed the provisions outlined in BS 5628 in terms of pre-stressed masonry (Wight 2006).

In New Zealand, the provision of pre-stressed masonry was appeared in 2004 edition of the New Zealand masonry standard, NZS 4230 'Design of reinforced Concrete Masonry Structures'.

2.2 Current State of Research

2.2.1 Out-of-Plane Versus In-Plane

The first study found in the literature on engineered use of pre-stressed masonry was carried out by Anderegg and Dalzell (1935). They conducted tests on masonry floor systems. A number of tests were carried out on masonry beams and columns and failures in compression and diagonal shear or tension were reported (Devalapura et al. 1997).

In 1953 in the United Kingdom, Samuely applied pre-stressing technique to brickwork piers in a school, and observed an increase in cracking moment capacity of the piers (Shrive 1988). Ramaswamy (1953) and Taylor (1961) applied pre-stressed technique to stone masonry walls.

During the mid 1960s, more research was conducted on pre-stressed masonry in England, Australia, and New Zealand (Ganz 1990; Geschwindner and Ostag 1990; Ungstad et al. 1990; Schultz and Scolforo 1991; Dawe and Aridru 1992; Bean and Schultz 2003; Ganz 2003).

In New Zealand, Hinkley (1966) applied post-tensioning to one-story brick masonry wall and concluded that vertical pre-stressing can improve the strength and the behavior of the walls. In Australia, Rosenhaupt et al. (1967) carried out a test on a concrete masonry wall and showed that the strength can be accurately predicted using truss analogy analysis. In England, Thomas (1969) and Mehta and Fincher (1970) carried out experimental tests on pre-stressed brick masonry beams and Hanion (1970) presented the application of post-tensioning techniques in retrofitting of several church steeples. Curtin and his co-investigators carried out several studies on pre-stressed masonry (Curtin 1986a, b, 1987; Curtin and Beck 1986; Curtin et al. 1975, 1982). They found that the bending resisting of brick wall system could be increased significantly by introducing pre-stressing.

Past research on post-tensioned masonry walls was more focused on the out-of-plane response of post-tensioned brick or block masonry walls, as presented in research review papers by Lissel et al. (2000) and Schultz and Scolforo (1991).

2.2.2   Research on In-Plane Behavior of Masonry

In-plane loading of an unbonded post-tensioned masonry wall leads to formation of a horizontal crack at the wall-footing interface. By increasing the lateral load, rocking of the wall governs the behavior, which is characterized by rotation about the wall's toe. This localizes the plastic region to the toe and enhances the drift capacity. More importantly, the wall can return to its original vertical alignment if sufficient residual pre-stress remains in the tendons and the tendons are not yielded (Wight and Ingham 2008).

An early study of the in-plane behavior of masonry walls was undertaken by Page and Huizer (1988). Using hollow clay units (nominal dimensions: 400 mm long, 100 mm high, 200 mm width), three walls with dimensions 3000 mm high × 2500 mm long × 200 mm thick were constructed and tested under in-plane monotonic load. The wall specifications were:

Wall I: Ungrouted, vertically and horizontally pre-stressed
Wall II: Ungrouted, post-tensioned vertically
Wall III: Grouted, vertically reinforced

The test setup is shown in Fig. 2.1a. For all walls, mortar type S, with a proportion of 1:0.5:4.5 cement:lime:sand by volume was used and all the pre-stressing rods were 15 mm Dywidag high strength rods. During the research the overall

(a)   **(b)**

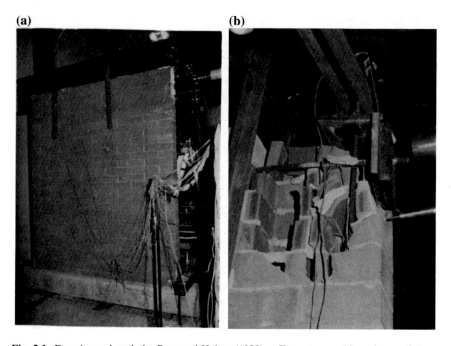

Fig. 2.1 Experimental study by Page and Huizer (1988). **a** Test setup, and **b** anchorage failure

Table 2.1 Experimental result of the walls tested by Page and Huizer (1988)

Wall	Load at decompression point (kN)	Ultimate strength (kN)	Failure mode
I	50	146	Premature local web splitting near point of load application
II	50	175	Diagonal tension failure
III	0	115	Diagonal tension failure

response of each wall was monitored and surface strains in critical locations and progressive cracking were compared. The authors concluded that post-tensioning was potentially an extremely effective method to increase the shear stiffness and rocking strength of a masonry wall. Although Wall I experienced a premature local failure as shown in Fig. 2.1b, the authors suggested that the potential benefit could be even more substantial if both horizontal and vertical pre-stressing was used to suppress the formation of diagonal tensile cracks in the tensile region of the wall (Table 2.1).

Rosenboom and Kowalsky (2004), investigated the in-plane behavior of post-tensioned walls under cyclic load by testing five large scale clay brick masonry walls (Fig. 2.2). The variables they focused on were bonded versus unbonded post-tensioning, confined versus unconfined masonry and grouted versus ungrouted masonry. According to their test results,

– To develop a rocking mechanism and a stable compression strut, the walls should be grouted.
– Walls without a grouted cavity do not allow a stable compression strut to develop and presented a brittle failure response.
– Unbonded post-tensioned masonry exhibited a low degree of energy dissipation. To overcome this, mild supplemental reinforcing can be incorporated to the base of the wall configurations. Although the displacement capacity remained unchanged, the strength of the wall increased significantly.
– Although, according to the test result, the unbonded post-tensioned wall dissipated less energy and should be designed for a higher level of base shear compared with the wall with mild steel, the benefit of reduced damage is significant and from performance based design it is more favorable.
– The wall with bonded PT bars added a degree of complexity to the construction, and since it exhibited a poor performance, it was not recommended.

Laursen and Ingham (2000a, b, 2004) undertook extensive research on unbonded post-tensioned concrete masonry walls. In phase I of their research they investigated the effects of aspect ratio, pre-stress level and grout infill through experimental testing of eight walls. They concluded that:

– Pre-stressed concrete masonry walls exhibit a nonlinear elastic behavior characterized by a rocking mechanism.

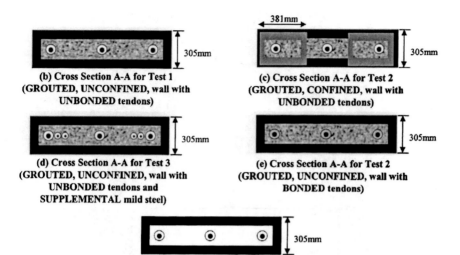

(b) Cross Section A-A for Test 1
(GROUTED, UNCONFINED, wall with
UNBONDED tendons)

305mm

381mm

(c) Cross Section A-A for Test 2
(GROUTED, CONFINED, wall with
UNBONDED tendons)

305mm

(d) Cross Section A-A for Test 3
(GROUTED, UNCONFINED, wall with
UNBONDED tendons and
SUPPLEMENTAL mild steel)

305mm

(e) Cross Section A-A for Test 2
(GROUTED, UNCONFINED, wall with
BONDED tendons)

305mm

(f) Cross Section A-A for Test 1 (UNGROUTED,
UNCONFINED, wall with UNBONDED tendons)

305mm

Fig. 2.2 Test setup and wall configuration tested by Rosenboom (2002)

- The level of energy dissipation in unbonded post-tensioned masonry is comparatively low.
- The self-centering behavior was retained even after tendon yielding.
- Partially grouted and ungrouted PT-MWs exhibited limited drift capacity and ductility and could fail in shear (Fig. 2.3).

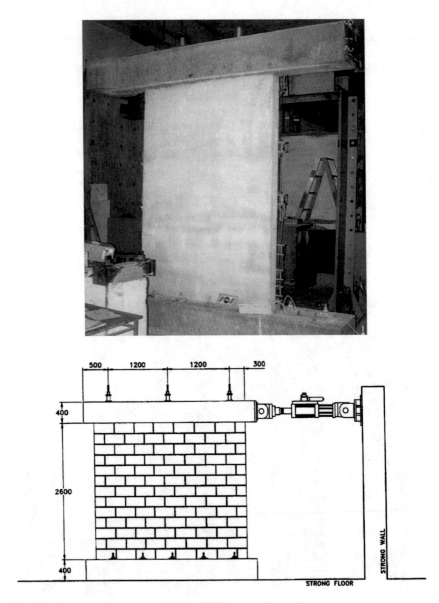

Fig. 2.3 Test setup (Laursen and Ingham 2000a)

In phase II of their study, Laursen and Ingham (2000b) tested PT-MWs (140 mm   thick × 3000 mm   length × 2800 mm   height)   by   incorporating strengthened masonry and enhanced energy dissipation. They concluded:

- Strengthening of the compression zones of unbonded PT-MWs can improve the wall displacement capacity and delay the onset of strength degradation.
- Use of confining plates can increase displacement capacity and reduce the wall damage in comparison with unconfined PT-MWs.
- In squat walls, losses can be expected due to yielding of the PT steel at relatively low wall drift ratios.
- Even after steel yielding, self-centering behavior is expected.

In phase III of their study, Laursen and Ingham (2004) tested two large scale 3-storey PT-MWs under in-plane cyclic loading and observed that:

- Strengthening the compression zones with confining plates can help the wall to withstand severe cyclic loading and high levels of drift.
- Relatively little energy dissipation was observed during cycling of the walls.
- Only localized damage occurred at the lowest 300–400 mm of the wall toe, which makes earthquake damage simple to repair (Fig. 2.4).

Wight (2006) conducted two sets of tests on unbonded PT-MWs. In phase I, he investigated the response of two partially grouted walls (Fig. 2.5a) and concluded that partially grouted PT-MWs with relatively low pre-stressing reinforcement ratios can sustain moderate levels of lateral displacement.

Shaking table tests of solid and perforated PT-MW specimens and simple square structures were also carried out in phase II of their experimental study (Wight et al.

Fig. 2.4 Test setup (Laursen and Ingham 2004)

Fig. 2.5 Test setup (Wight 2006)

2006; Wight et al. 2007) (Fig. 2.5b). Through this study they verified the ability of such walls to return to their original vertical alignment and withstand large numbers of excitations with minimal damage. The damage of the simple structure was consistent with that obtained for perforated walls, being bond beam cracking and cracking below the openings (Wight et al. 2007). Moreover Wight and Ingham (2008) concluded that reducing the level of the initial PT force significantly increased the wall displacement capacity.

Ewing (2008) conducted experimental tests on PT-MWs with openings and concluded that it is possible to design perforated unbonded post-tensioned clay brick masonry walls to maintain all of the benefits of solid PT-MWs, provided properly designed cold joints are included that divide a single unbonded perforated PT-MW into multiple piers. They also concluded that the size and location of the opening has a major effect on the overall response of the wall (Ewing 2008) (Fig. 2.6).

In a very recent study, Ryu et al. (2014) conducted a finite element analysis to investigate the effect of PT bar spacing on the behavior of unbonded PT-MWs. In-plane behavior of unbonded PT-MWs made up of clay brick masonry was investigated in their study. Experimental results of prism tests were used to calibrate the material parameters of the numerical model. Homogenous isotropic material was considered to simulate masonry using smeared-crack (macro) modeling. To validate the finite element model, the experimental results of the two walls tested by Rosenboom and Kowalsky (2004) and Ewing (2008) were used. It was shown that

Fig. 2.6 Test setup and result (Ewing 2008)

the developed FE model could accurately capture the behavior of the tested
PT-MWs. The numerical model was also successful in determining the failure mode
as well as the force in the PT tendons. The verified FE model was then used to carry
out parametric study to investigate the effects of various parameters including the
tendon spacing, aspect ratio, and shear reinforcement on the response of PT-MWs.
Based on the results of the parametric study, they concluded that:

- For walls with PT bar spacing of less than 2 m, ignoring the elongation of the
  bar in the flexural strength calculation provides a very conservative design.
- For walls with PT bar spacing of less than 2 m, adding horizontal shear reinforcement does not improve the strength of the walls. However, for walls having
  wide spaced PT bars, adding a small amount of horizontal shear reinforcement
  significantly increases the shear strength and improves the performance.
- Displacement ductility of the walls decreases with an increase in the tendon
  spacing.

2.3   Gap in Knowledge

This chapter consists of a literature review of research on unbonded post-tensioned
masonry walls. The experimental work found in the literature included tests conducted by Page and Huizer (1988), Rosenboom and Kowalsky (2004), Laursen and

Ingham (2000a, b, 2004), Wight (2006), Ewing (2008). In an early study conducted by Page and Huizer (1988), the authors concluded that post-tensioning is potentially an extremely effective method to increase the shear stiffness and rocking strength of a masonry wall. Analysing the results of tests on five post-tensioned clay masonry walls, undertaken by Rosenboom and Kowalsky (2004), they reported that to develop a rocking mechanism, the wall should be grouted. They also concluded that unbonded post-tensioned masonry exhibited a low degree of energy dissipation. In a series of experimental studies on post-tensioned concrete masonry walls, Laursen and Ingham (2000a, b, 2004) indicated that use of confining plates can increase displacement capacity and reduce the wall damage in comparison with unconfined PT-MWs. Moreover they reported that in squat walls, losses can be expected due to yielding of the PT steel at relatively low wall drift ratios. Conducting shaking table tests of solid and perforated PT-MW specimens and simple square structures, Wight et al. (2006, 2007) verified the ability of such walls to return to their original vertical alignment and withstand large numbers of excitations with minimal damage. Moreover, they concluded that reducing the level of the initial PT force significantly increased the wall displacement capacity. According to the results of tests on PT-MWs with openings, Ewing (2008) showed the size and location of the opening has a major effect on the overall response of the PT-MWs. Conducting finite element analysis of PT-MWs, Ryu et al. (2014) reported that for walls with PT bar spacing of less than 2 m, adding horizontal shear reinforcement does not improve the strength of the walls. However, for walls having wide spaced PT bars, adding a small amount of horizontal shear reinforcement significantly increases the shear strength and improves the performance.

To the best knowledge of the author, no study has yet investigated the seismic parameter of PT-MWs. Moreover, the rocking mechanism of the PT-MWs is not well understood. This study aims to get a better understanding of the seismic behavior of PT-MWs and to determine seismic response factors including ductility, R and c_d for such walls. Another objective of this thesis is to develop expressions and guidelines for unbonded PT-MWs and to incorporate an analytical approach to determine the in-plane force-displacement response of PT-MWs. An experimental study was also conducted on four post-tensioned concrete masonry walls to verify the expressions and guidelines presented in the study.

References

Anderegg FO, Dalzell C L (1935) Pre-stressed ceramic members. In: Proceedings of the American society for testing and materials, American society for civil engineers (ASCE), vol 35, pt 2, pp 447–456
AS3700 (2011) "Masonry Structures", Australian Standard, Sydney, NSW, Australia
Bean J, Schultz A (2003) Flexural capacity of post-tensioned masonry walls: code review and recommended procedure. PTI J 1:28–44
Curtin WG (1986a) An investigation of the structural behavior of post-tensioned brick diaphragm walls. Struct Eng 64B(4):77–84

Curtin WG (1986b) Post-tensioning opens brickwork frontiers. Contract J 333(558):26–27

Curtin WG (1987) Site testing and lab research for the design development of prestressed brickwork. In: Practical design of masonry structures. Thomas Telford Ltd., London, pp 237–254

Curtin WG, Adams S, Sloan M (1975) The use of post-tensioned brickwork in the SCD system. In: Proceedings of the British Ceramic Society, Load-Bearing Brickwork (5), no 24, pp 233–245

Curtin WG, Shaw G, Beck JK, Pope LS (1982) Post-tensioned free cantilever diaphragm wall project. In: Reinforced and prestressed masonry. The Institution of Civil Engineers, Thomas Telford, Ltd., London, pp 79–88

Curtin WG, Beck JK (1986) The structural design of a concrete blockwork retaining wall. Concrete 20(7):19–21

Dawe JL, Aridru GG (1992) Post-tensioned concrete masonry walls subjected to uniform lateral loading. In: Proceedings of the 6th Canadian masonry symposium, University of Saskatchewan, Saskatoon, Saskatchewan, pp 201–212

Devalapura RK, Krause GL, Sweeney SC, Littler D, Staab E (1997) Construction productivity advancement research (CPAR) program. Development of an innovative post-tensioning system for prestressed clay brick masonry walls. DTIC Document

Ewing B (2008) Performance of post-tensioned clay brick masonry walls with openings. Ph.D., thesis, North Carolina State University, Raleigh, NC, USA

Ganz HR (1990) Post-tensioned masonry structures. In: SL report series 2. VSL International Ltd., Berne, Switzerland

Ganz HR (2003) Post-tensioned masonry around the world. Concr Int-Detroit 25:65–70

Geschwindner LF, Ostag WP (1990) Post-tensioned single-wythe concrete masonry walls. In: Proceedings of the 5th North American masonry conference, pp 1123–1134

Hanion JRG (1970) Concrete masonry in New Zealand: prestressed concrete masonry. Concrete 4 (9):356–358 (The Concrete Society, England)

Hinkley AT (1966) Tests of one-story prestressed brickwork shear walls. NZ Eng 21(6):245–252

Laursen PPT (2002) Seismic analysis and design of post-tensioned concrete masonry walls. Ph.D., thesis, University of Auckland, Department of Civil and Environmental Engineering, Auckland, New Zealand

Laursen P, Ingham J (2000a) Cyclic in-plane structural testing of pre-stressed concrete masonry walls, volume A: evaluation of wall structural performance. Report No. 599, School of Engineering, Department of Civil and Resources Engineering, University of Auckland, Auckland, New Zealand

Laursen P, Ingham J (2000b) Cyclic in-plane structural testing of pre-stressed concrete masonry walls, volume B: data resource. Report No. 600, School of Engineering, Department of Civil and Resources Engineering, University of Auckland, Auckland, New Zealand

Laursen P, Ingham J (2004) Structural testing of large-scale post-tensioned concrete masonry walls. J Struct Eng 130(10):1497–1505

Lissel SL, Shrive NG, Page AW (2000) Shear in plain, bed joint reinforced, and post-tensioned masonry. Can J Civ Eng 27(5):1021–1030

Mehta KC, Fincher D (1970) Structural behavior of pre-tensioned prestressed masonry beams. In: Proceedings of the 2nd international brick masonry conference (SIBMAC), Stoke-on-Trent, England, pp 215–219

MSJC (2005) Building code requirements for masonry structures. The Masonry Society (TMS), ACI 530/ASCE 5, TMS 402, American Concrete Institute, Detroit

Page AW (2001) Pre-stressed masonry-recent Australian research and code provisions. Structures Congress and Exposition, Washington, D.C.

Page A, Huizer A (1988) Racking behavior of pre-stressed and reinforced hollow masonry walls. Mason Int 2(3):97–102

Ramaswamy GS (1953) Prestressing and assembly of stone blocks by post-tensioning. Indian Concr J 27(12):450–451

Rosenboom OA (2002) Post-tensioned clay brick masonry walls for modular housing in seismic regions. M.S. thesis, North Carolina State University, Raleigh, NC, USA

Rosenboom OA, Kowalsky MJ (2004) Reversed in-plane cyclic behavior of post-tensioned clay brick masonry walls. J Struct Eng 130(5):787–798

Rosenhaupt S, Beresford FD, Blakey FA (1967) Test of a post-tensioned concrete masonry wall. ACI J Proc 64(12):829–837

Ryu D, Wijeyewickrema A, ElGawady M, Madurapperuma M (2014) Effects of tendon spacing on in-plane behavior of post-tensioned masonry walls. J Struct Eng 140(4):04013096

Schultz AE, Scolforo MJ (1991) Overview of pre-stressed masonry. Mason Soc J 10(1):6–21

Shrive NG (1988) Post-tensioned masonry—status and prospects. In: Society for Civil Engineering Annual Conference, Calgery, Canada, May 25–27, pp 679–696

Taylor JB (1961) Prestressed granite masonry for a retaining wall. Proc Am Soc Civ Eng 31 (1):33–34

Thomas K (1969) Current post-tensioned and prestressed brickwork and ceramics in Great Britain. In: Johnson FB (ed) Proceedings of the international conference on masonry structural systems. Gulf Publishing Co., pp 285–301

Ungstad DG, Hatzinikolas MA, Warwaruk J (1990) Prestressed concrete masonry walls. In: Proceedings of the 5th North American Masonry Conference, University of Illinois at Urbana Champaign, Champaign, IL, pp 1147–1161

Wight GD (2006) Seismic performance of a post-tensioned concrete masonry wall system. Ph.D. dissertation, Department of Civil and Environmental Engineering, University of Auckland, Auckland, New Zealand

Wight GD, Ingham JM (2008) Tendon stress in unbonded post-tensioned masonry walls at nominal in-plane strength. J Struct Eng 134(6):938–946

Wight GD, Ingham JM, Kowalsky MJ (2006) Shaketable testing of rectangular post-tensioned concrete masonry walls. ACI Struct J 103(4):587

Wight GD, Kowalsky M, Ingham J (2007) Shake table testing of post-tensioned concrete masonry walls with openings. J Struct Eng 133(SPECIAL ISSUE: Precast pre-stressed concrete structures under natural and human-made hazards):1551–1559

Chapter 3
Strength and Seismic Performance Factors of Post-tensioned Masonry Walls

In this chapter, the behavior of PT-MWs is investigated using a database of tested walls. The accuracy of ignoring elongation of PT bars which is considered in the current masonry standard joint committee (MSJC 2013) code in evaluating the strength of PT-MWs is studied using the available test results. Using the experimental results, the structural response parameters including ductility, response modification factor and displacement amplification factor are determined for different types of walls including fully grouted, partially grouted, ungrouted walls, walls with confinement plates, walls with supplemental mild steel and walls with an opening.

3.1 Introduction

One of the earliest studies on post-tensioned masonry walls (PT-MWs) was reported by Samuely in 1953 in the United Kingdom, who examined post-tensioned brickwork piers in a school (Shrive 1988). Since then, the applications of post-tensioned masonry have increased with more emphasis on the out-of-plane response (Schultz and Scolforo 1991; Laursen 2002; Lissel and Shrive 2003; Bean 2007). One of the earliest experimental studies on in-plane behavior of PT-MWs was carried out by Page and Huizer (1988). Three PT-MWs were tested under in-plane monotonic load and it was concluded that post-tensioning was potentially an effective method to increase the strength of masonry walls.

Under in-plane loading of an unbonded post-tensioned masonry wall, wall cracks form at the wall-footing interface. By increasing the in-plane load, rocking of the wall occurs, which is characterized by rotation about the wall's toe. This

With permission from ASCE: Hassanli R., ElGawady M.A. and Mills J.E., Strength and seismic performance factors of post-tensioned masonry walls, Journal of Structural Engineering, 141 (11), 2015.

localizes the damage to the toe region. More importantly, the wall can return to its original vertical alignment if sufficient residual post-tensioning force remains in the tendons and the tendons do not develop significant inelastic strains (Wight and Ingham 2008).

The main drawback of unbonded post-tensioned systems is that the energy dissipation is comparatively low compared to conventional reinforced systems (ElGawady and Sha'lan 2011; ElGawady et al. 2010; Erkmen and Schultz 2009; Laursen 2002; Rosenboom 2002; Wight 2006). To improve the behavior of unbonded PT-MWs, different methods have been tried, including incorporating supplemental mild steel or high strength concrete blocks. Rosenboom (2002) incorporated supplemental mild reinforcing steel between a wall and its footing. While this increased the strength, the displacement capacity slightly decreased.

Incorporating confinement plates at the toe region increases the masonry ultimate strain capacity and hence the displacement capacity of the wall (Rosenboom 2002). However, it does not increase the energy dissipation (Laursen 2002). The differences between the behavior of bonded and unbonded post-tensioned walls were also investigated (Rosenboom 2002). Bonded PT bars added a degree of complexity to the construction; however, bonded post-tensioned walls exhibited poor performance and hence were not recommended for construction (Rosenboom 2002).

Ungrouted, partially grouted, and fully grouted PT-MWs exhibit different behavior, and failure mechanisms. Unlike fully grouted walls, partially grouted and ungrouted wall specimens exhibited a limited drift capacity and ductility and mainly failed in shear (Laursen 2002). Partially grouted walls do not allow a stable compression strut to be formed (Rosenboom 2002).

Experimental tests have also been carried out on perforated PT-MWs (Wight 2006; Ewing 2008). Ewing (2008) concluded that it is possible to design perforated unbonded post-tensioned clay brick masonry walls to maintain all of the benefits of solid PT-MWs, provided properly designed cold joints are included that divide a single unbonded perforated PT-MW into multiple piers. Shaking table tests of solid and perforated PT-MW specimens and simple square structures were also carried out by Wight (2006). This study verified the ability of such walls to return to their original vertical alignment and withstand large numbers of excitations with minimal damage. The damage of the simple structure was consistent with that obtained for perforated walls, being bond beam cracking and cracking below the openings (Wight 2006).

The current strength expressions developed for PT-MWs either incorporate simplification, as with the approach of MSJC (2013), where PT forces are considered constant during dynamic excitation, or they are based on semi-empirical equations (Ryu et al. 2014; Wight 2006).

This study aims to get a better understanding of the seismic behavior of PT-MWs. To the best knowledge of the authors, there is no single integrated study to evaluate the ability of the current code expressions in predicting the strength of

PT-MWs according to the available test results. Moreover, there has been no study to determine the seismic response factors (i.e. R and c_d) of PT-MWs. In terms of seismic response factors, the MSJC (2013) considers unbonded PT-MWs as unreinforced masonry, resulting in a comparatively small value for response modification factor and displacement amplification factor; based on the assumption that the walls essentially remain elastic. The test results of 31 PT-MWs available in the literature were collected and examined. The collected results were used to investigate the accuracy of ignoring PT bar elongation in MSJC (2013) in predicting the lateral strength of PT-MWs. Moreover, the seismic response factors including ductility, R and c_d were determined for these specimens and compared with those in ASCE 7 (2010) and MSJC (2013).

3.2   Test Database

Following a comprehensive literature review of PT-MWs a database of 31 specimens has been collected. The database includes Laursen's tests comprising eight wall specimens (L1-Wall5 to L1-Wall) from series 1, five from series 2 (L2-Wall1 to L2-Wall5) and two from series 3 (L3-Wall1 to L3-Wall2) (Laursen 2002); five walls tested by Rosenboom (2002) (R-Wall1 to R-Wall5); three by Ewing (2008) (E-Wall1 to E-Wall3); two from series 1 of tests (W1-Wall1 to W1-Wall2) and six from series 2 (W2-WallD1 to W2-WallD6) by Wight (2006). The configurations of the walls are as follows (Table 3.1):

- *Fully grouted unbonded masonry walls*: Seven walls tested under cyclic load, including walls L1-Wall1 to L1-Wall6 and R1-Wall1.
- *Partially grouted or ungrouted masonry walls*: Seven partially grouted walls and two ungrouted post-tensioned wall specimens in the test-database tested under cyclic load including L1-Wall7 to L1-Wall8, W1-Wall1, W1-Wall2, R1-Wall5, W1-WallD1 and W1-WallD4. The series 2 of tests conducted by Wight (2006) is not considered here to calculate the structural response factors as for most of the specimens, testing stopped due to an inability to generate larger ground accelerations because of the limited capacity of the shaking table.
- *Walls with confinement plates or supplemental mild steel*: Eleven walls, including walls R-Wall2, R-Wall3, L2-Wall1 to L2-Wall5, L3-Wall1, L3-Wall2, E-Wall2 and E-Wall3, in which confinement plates or supplemental mild steel were incorporated to enhance the behavior of the system. The last two walls from Ewing were categorized in the wall-with opening group.
- *Bonded masonry walls*: One specimen, R-Wall4.
- *Walls with openings*: Five walls, E-Wall1 to E-Wall3, which were tested under cyclic load and W2-WallD5 and W2-WallD6 which were tested under dynamic load.

Table 3.1 Post-tensioned masonry wall database

Reference	Wall designation	Original designation	Specification	Material unit	Loading
Laursen (2002)	L1-Wall1	FG:L3.0-W20-P3	FG	CMU	Cyclic
	L1-Wall2	FG:L3.0-W15-P3	FG	CMU	Cyclic
	L1-Wall3	FG:L3.0-W15-P2C	FG	CMU	Cyclic
	L1-Wall4	FG:L3.0-W15-P2E	FG	CMU	Cyclic
	L1-Wall5	FG:L1.8-W15-P2	FG	CMU	Cyclic
	L1-Wall6	FG:L1.8-W15-P3	FG	CMU	Cyclic
	L1-Wall7	PG:L3.0-W15-P2	PG	CMU	Cyclic
	L1-Wall8	UG:L1.8-W10-P2	Ungrouted	CMU	Cyclic
	L3-Wall1	S3-1	FG + confinement plate	CMU	Cyclic
	L3-Wall2	S3-2	FG + confinement plate	CMU	Cyclic
	L2-Wall1	FG:L3.0-W15-P1-CP	FG + confinement plate	CMU	Cyclic
	L2-Wall2	FG:L3.0-W15-P2-CP	FG + confinement plate	CMU	Cyclic
	L2-Wall3	FG:L3.0-W15-P2-CP-CA	FG + confinement plate + constant axial load	CMU	Cyclic
	L2-Wall4	FG: L3.0-W15-P2-CP-CA-ED	FG + confinement plate + constant axial load + energy dissipation	CMU	Cyclic
	L2-Wall5	FG:L3.0-W15-P2-HB	FG + confinement plate + high strength block	CMU	Cyclic
Rosenboom (2002)	R-Wall1	Test1	FG	Brick	Cyclic
	R-Wall2	Test3	FG + confinement plate	Brick	Cyclic
	R-Wall3	Test2	FG + supplemental mild steel	Brick	Cyclic
	R-Wall4	Test5	FG + bonded Tendon	Brick	Cyclic
	R-Wall5	Test4	Ungrouted	Brick	Cyclic
Wight (2006)	W1-Wall1	C1	PG	CMU	Cyclic
	W1-Wall2	C2	PG	CMU	Cyclic

(continued)

Table 3.1 (continued)

Reference	Wall designation	Original designation	Specification	Material unit	Loading
		D1	PG	CMU	Dynamic
	W2-WallD2	D2	PG	CMU	Dynamic
	W2-WallD3	D3	FG	CMU	Dynamic
	W2-WallD4	D4	PG + control joint	CMU	Dynamic
	W2-WallD5	D5	With door opening (opening size: 813 * 2032)	CMU	Dynamic
	W2-WallD6	D6	With window opening (opening size: 813 * 1219)	CMU	Dynamic
Ewing (2008)	E-Wall1	Wall1	With door opening (opening size: 900 * 1520)	Brick	Cyclic
	E-Wall2	Wall2	With door opening (opening size: 900 * 1520) + confinement plate	Brick	Cyclic
	E-Wall3	Wall3	With door opening (opening size: 900 * 1520) + confinement plate + horizontal steel	Brick	Cyclic

FG fully grouted, *PG* partially grouted, *CMU* concrete masonry unit

3.3 Prediction of In-Plane Shear Strength of Unbonded PT-MWs

MSJC (2013) has no procedure for estimating f_{ps} for unbonded post-tensioned masonry shear walls. According to MSJC (2013), instead of a more accurate determination, f_{ps} for members with unbonded pre-stressing bars can conservatively be taken as f_{se}. To investigate the accuracy of this approach in predicting the strength of PT-MWs based on the available test database, the base shear from the flexural expression (Eq. 3.1) and from the shear expression (Eq. 3.3) are calculated and presented in Table 3.2. The ultimate base shear recorded in the test is also provided in the table.

$$M_n = \sum f_{se}A_{ps}\left(d - \frac{a}{2}\right) + N\left(\frac{l_w}{2} + \frac{a}{2}\right) \tag{3.1}$$

$$\text{where: } a = \frac{\sum f_{se}A_{ps} + N}{0.8t_w f'_m} \tag{3.2}$$

where f_{se} is the effective stress in the PT bar after immediate stress losses, A_{ps} is the area of the PT bar, N is the gravity load, f'_m is the compressive strength of masonry, a is the depth of the equivalent compression zone, d is the distance from extreme compression fiber to centroid of tension reinforcement, t_w is the thickness and l_w is the length of the wall.

For PT-MWs having no bonded reinforcement, the shear strength can be calculated as follows:

$$V_n = min\begin{cases} 0.315A_n\sqrt{f'_m} & (a) \\ 2.07A_n & (b) \\ 0.621A_n + 0.45N & (c) \end{cases} \tag{3.3}$$

And for reinforced masonry walls, the nominal shear strength, V_n, is the sum of a masonry component (V_m), a reinforcement component (V_s) and the effect of axial stress (V_p):

$$V_n = V_m + V_s + V_p \tag{3.4}$$

where

$$V_m = 0.083\left[4.0 - 1.75\left(\frac{M}{Vl_w}\right)\right]A_n\sqrt{f'_m}, \tag{3.5}$$

$$V_p = 0.25P_u \tag{3.6}$$

Table 3.2 Failure mode, calculated and measured strengths

Wall	Specification	Failure mode	Test results V_{Exp} (kN)	Flexural expression V (kN)	Shear expression V (kN)
L1-Wall1	FG[a]	Flexural	561.0	325.7	637.5
L1-Wall2	FG	Flexural	465.0	361.0	514.1
L1-Wall3	FG	Flexural	373.0	345.9	559.2
L1-Wall4	FG	Flexural	373.0	342.8	558.2
L1-Wall5	FG	Flexural	178.0	145.8	359.4
L1-Wall6	FG	Shear	266.0	215.4	340.5
R-Wall1	FG	Compression strut failure	330.9	214.0	552.3
L3-Wall1	FG + confinement plate	Masonry crushing-gradual flexural failure	212.0	209.8	447.8
L3-Wall2	FG + confinement plate	Masonry crushing-gradual flexural failure	165.0	148.6	396.0
L1-Wall7	PG[a]	Shear	120.0	99.2	206.8
L1-Wall8	Ungrouted	Shear	100.0	69.4	112.2
R-Wall2	FG + confined plate	Flexural	347.1	214.2	591.9
R-Wall3	FG + supplemental mild steel	Flexural	365.6	248.7	591.9
R-Wall4	FG + bonded Tendon	Flexural	340.7	304.1	591.9
R-Wall5	Ungrouted	Shear	242.1	196.6	321.1
L2-Wall1	FG + confined plate	Masonry crushing, flexural failure	249.0	185.2	413.8
L2-Wall2	FG + confined plate	Masonry crushing, flexural failure, diagonal crack	395.0	316.2	514.1
L2-Wall3	FG + confined plate + constant axial load[b]	Masonry crushing, flexural failure	338.0	329.9	543.9
L2-Wall4	FG + confined plate + constant axial load +energy dissipation[c]	Masonry crushing, flexural failure	414.0	332.0	551.1
L2-Wall5	FG + confined plate + high strength block	Masonry crushing, flexural failure, diagonal crack	380.0	320.1	467.8
W1-Wall1	PG	Shear-flexural	105.6	76.3	196.5
W1-Wall2	PG	Anchorage	46.4	29.2	127.7

[a]*FG* fully grouted, *PG* partially grouted
[b]The force in the PT bars was kept constant during the test
[c]'Dog-bone'-type dampers were incorporated in the wall

$$V_s = 0.5 \left(\frac{A_v}{s_h} \right) f_{py} l_w \tag{3.7}$$

$$V_n \leq 0.50 A_n \sqrt{f'_m} \quad for \quad M/Vl_w \leq 0.25 \tag{3.8}$$

$$V_n \leq 0.33 A_n \sqrt{f'_m} \quad for \quad M/Vl_w \geq 1.00 \tag{3.9}$$

where A_n is net cross sectional area of the masonry wall, f_{py} is yield strength, s_h is spacing and A_v is area of the shear reinforcement. For post-tensioned walls, the axial post-tensioning force shall be included in determining P_u in the V_p component.

The mode of failure of the walls is also provided in Table 3.2. Flexural failure is characterized by yielding of the PT bars and/or masonry crushing at the toe of the wall. A diagonal crack in the wall signifies a shear failure. Shear-flexural failure is a combination of flexural and shear failures. As shown in Table 3.2, in all fully grouted walls except L1-Wall6, a flexural mode of failure governed the mechanism. However, none of the partially grouted or ungrouted walls included in the database exhibited a flexural failure mode. Comparing the experimental results with predicted values from MSJC's (2013) approach reveals that these expressions tend to underestimate the flexural strength of PT-MWs, due to ignoring the elongation of PT bars. For fully grouted walls, the maximum, minimum, and average ratios of V_{exp}/V_{calc} were found to be equal to 0.99, 0.58, and 0.8, respectively, implying a relatively conservative prediction.

Figure 3.1 shows the effect of initial to yield stress ratio of the PT bars, f_i/f_{py}, on the value of V_{exp}/V_{calc} of the fully grouted walls, with and without confinement plates. As shown in the figure, as f_i/f_{py} increases the value of V_{exp}/V_{calc} increases and approaches a value of one. In other words, for smaller ranges of f_i/f_{py}, MJSC (2013) provides a very conservative prediction due to ignoring the elongation in the post-tensioning bars. For higher values of f_i/f_{py}, ignoring this elongation is justified.

Figure 3.2 shows the effect of bar spacing on the value of V_{exp}/V_{calc} in fully grouted walls without confinement plates. As shown in the figure, the database has

Fig. 3.1 Effect of f_i/f_{py} on V_{exp}/V_{cal}

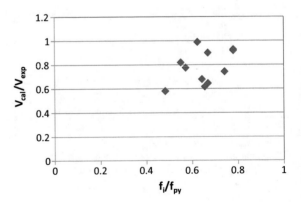

Fig. 3.2 Effect of spacing on V_{exp}/V_{cal}

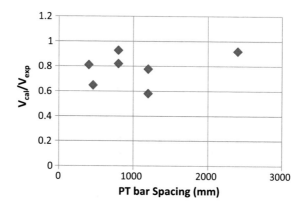

six specimens having small spacing (less than 1.2 m) between the bars. For this set of data, bar spacing had no significant effect on V_{exp}/V_{calc} with an average V_{exp}/V_{calc} of 0.76. For larger spacing between bars, the database includes only one specimen having spacing of 2.4 m. This one specimen has V_{exp}/V_{calc} of 0.92. It is worth noting that Ryu et al. (2014), using an extensive finite element analysis of post-tensioned masonry walls, found that for bar spacing smaller than 2.5 m, the average V_{exp}/V_{calc} was approximately 0.70. However, for larger spacing between bars, the average V_{exp}/V_{calc} approached 1.0. Hence, based on the analysis of the database and the Ryu et al. findings, it seems by ignoring the PT bar elongation in MSJC (2013), strength predictions are quite conservative for smaller spacing between bars and less conservative for larger spacing between bars. However, more experimental data is still required for larger spacing between bars.

As presented in Table 3.2, all of the partially grouted and ungrouted walls exhibited either a shear, shear/flexural, or anchorage mode of failure. None of these walls included either horizontal or vertical bonded steel; hence, according to MSJC (2013) they are considered as unreinforced masonry walls and Eq. 3.3 is applied. While all of these walls failed in shear related modes, ignoring the elongation of bars results in a flexural failure prediction. Therefore, MSJC (2013) over predicted the shear strength of partially grouted and ungrouted walls. Disregarding the wall W1-Wall2 which is not considered due to anchorage failure, the shear strength is over-predicted by 12–86%. More tests are required to verify the result provided here and to investigate if the MSJC (2013) expressions over predict the shear strength of ungrouted and partially grouted PT-MWs. Table 3.3 presents the detailed shear strength prediction of partially grouted and ungrouted walls. According to Eq. 3.3 the minimum of $0.315A_n \sqrt{f'_m}$, $2.07A_n$ and $0.621A_n + 0.45N$ controls the shear strength. Comparing $0.315A_n \sqrt{f'_m}$ and $2.07A_n$ indicates that the latter equation is not controlling unless $f'_m \geq 43.2$ MPa which is significantly higher than typical values of f'_m. As shown in Table 3.3, the equations $0.315A_n \sqrt{f'_m}$ and $0.621A_n + 0.45N$ control for ungrouted and partially grouted walls, respectively.

Table 3.3 Strength of ungrouted/partially grouted walls: shear expression

| Wall | Tested by | Specification | V_{exp} (kN) | Shear strength (kN) | | | V_{calc} (kN) | V_{calc}/V_{exp} |
				$0.315A_n\sqrt{f'_m}$	$2.07A_n$	$0.621A_n + 0.45N_u$		
L1-Wall7	Laurson-Set1-Wall7	PG	120.0	253.0	422.3	206.8	206.8	1.72
W1-Wall1	Wight-Set1-Wall1	PG	105.6	255.3	418.1	196.5	196.5	1.86
W1-Wall2	Wight-Set1-Wall2	PG	46.4	177.4	298.1	127.7	127.7	2.75
L1-Wall8	Laurson-Set1-Wall8	Ungrouted	100.0	112.2	190.4	176.8	112.2	1.12
R-Wall5	Rosenboom-Wall5	Ungrouted	242.1	392.5	510.8	607.1	392.5	1.62

3.4 Prediction of In-Plane Strength of Bonded PT-MWs

The only bonded PT-MW in the database is the one tested by Rosenboom and Kowalsky (2004). The wall configuration is presented in Table 3.4. Based on the strain compatibility method and equilibrium, the strength of the wall can be determined through the following process:

$$\varepsilon_{in,m} = \frac{post\ tensioning\ force}{E_m l_w t_w} \tag{3.10}$$

$$\varepsilon_{in,p} = \frac{\sigma_{in,p}}{E_p} \tag{3.11}$$

$$\varepsilon_{in,m} + \Delta\varepsilon_m = \varepsilon_{mu} \tag{3.12}$$

$$\Delta\varepsilon_{pi} = \left(\frac{d_i}{c} - 1\right) * \Delta\varepsilon_m \tag{3.13}$$

$$F_{si} = \left(\varepsilon_{in,p} + \Delta\varepsilon_{pi}\right) E_p A_{pi} \leq f_{py} A_{pi} \tag{3.14}$$

$$\sum F_{si} = 0.64 f'_m t_w c \tag{3.15}$$

where $\varepsilon_{in,m}$ and $\varepsilon_{in,p}$ are the strain values in the masonry and PT bar respectively, due to initial pre-stressing, $\Delta\varepsilon_{pi}$ and $\Delta\varepsilon_{pi}$ are incremental strains in the masonry and PT bar due to wall rotation at the base, f_{py} is the yield strength and E_p is the elastic modulus of the PT bar, c is the compression zone length, a is the depth of the constant stress block, and F_{si} is the developed force in the ith PT bar. The unknown parameter, c, can be determined through an iterative process. Table 3.5 presents the strain in the steel and the corresponding moment capacity obtained from the strain compatibility method. As presented in the table, the moment capacity of the wall is 742 kNm corresponding to a strength of 304.1 kN, which is approximately 10% less than 340.7 kN, the reported experimental base shear. To the best knowledge of the authors, this is the only recorded test carried out on a bonded post-tensioned masonry wall. According to this test result, the prediction of lateral strength using the strain compatibility method, which is recommended by masonry standards, provided an acceptable estimation.

3.5 Bilinear Idealization of Capacity Curves

The bilinear approximation of the capacity curve is used to determine the system seismic parameters, R and c_d. Approximation of the transition point of the elasto-plastic behavior (pseudo yield) can be obtained by creating a bilinear idealization of the capacity curve of the system following the Applied Technology

Table 3.4 Configuration of R-Wall4

Wall	Specification	h (mm)	t (mm)	l_w (mm)	f'_m (MPa)	No. of bars	A_{pi} (mm^2)	f_{py} (MPa)	Initial stress (MPa)	Precompression force (kN)
R-Wall4	FG + bonded	2440	305	1220	25.5	3	550	890	509	840

Table 3.5 Moment calculation using compatibility method

c (mm)	d	$\varepsilon_{in,m}$	$\varepsilon_{in,p}$	$\Delta\varepsilon_m$	$\Delta\varepsilon_s$	F_i (kN)	C (kN)	M (kNm)
227.6	152	0.00013	0.00248	0.00337	−0.00112	154	1133	9
	610	0.00013	0.00248	0.00337	0.00567	490		254
	1068	0.00013	0.00248	0.00337	0.01246	490		478
Total						1133		742

Council ATC-40 (1996) procedure and using the equal energy method. To generate the bilinear curve, a maximum displacement (Δ_{max}) needs to be considered, which can be taken as the point when the lateral strength degrades by 20%. It is worth noting that, unlike structural elements that have bonded reinforcement, the pseudo yielding point in the idealized backbone curve corresponds to the point where significant nonlinearity occurs. In unbonded post-tensioned masonry walls, the nonlinearity occurs when the interface joint at the base of the wall significantly opens leading to stiffness softening. Hence, the pseudo yielding point does not necessarily correspond to yielding of any bars, but rather it corresponds to stiffness softening in the wall. Examples of bilinear approximations of the load displacement curve of the tested specimens are shown in Fig. 3.3a.

Table 3.6 presents the parameters determined from bilinear elasto-plastic curves of the test specimens. In this table, Δ_y and f_y are the yield displacement and yield strength respectively and μ is the displacement ductility defined as Δ_{max}/Δ_y. The ductility values for fully grouted PT-MWs vary between 4.8 and 34.5, as presented in Table 3.6. The ductility increased and reached a value of 98.4 when confinement plates were used. Ungrouted PT-MWs exhibited the least ductility, with values ranging from 3.5 to 5.3. Partially grouted walls displayed ductility values that varied between 9.9 and 13.6.

Figure 3.4 presents the ductility versus axial stress ratio. Axial stress ratio is defined as the total applied axial load, including the initial post-tensioning load, divided by the net cross sectional area, i.e. f_m, normalized by f'_m. While the trend lines in Fig. 3.4 indicate that ductility is dependent on the axial stress ratio, f_m/f'_m, the number of test specimens in each category is limited and hence more test specimens are required before a firm conclusion can be drawn. For example, the numbers of ungrouted and partially grouted test specimens are only 2 and 3, respectively, while the numbers of test specimens increased to 7 and 8 specimens for fully grouted and specimens having confinement plates, respectively. For specimens with confinement plates or fully grouted the trend is the same, with R^2 values of 0.33 and 0.17, respectively. While it is not a strong correlation, all subsets of test specimens indicate that ductility decreases with increasing f_m/f'_m. Moreover, Laursen (2002) tested two identical walls, walls L1-Wall5 and L1-Wall6, except that the walls were subjected to two different levels of post-tensioning force. The second wall, wall L1-Wall6 was subjected to f_m/f'_m of 0.17 and displayed a brittle failure characterized by diagonal cracking, due to tensile splitting of the masonry compression strut formed between the post-tensioning anchorage of the loading

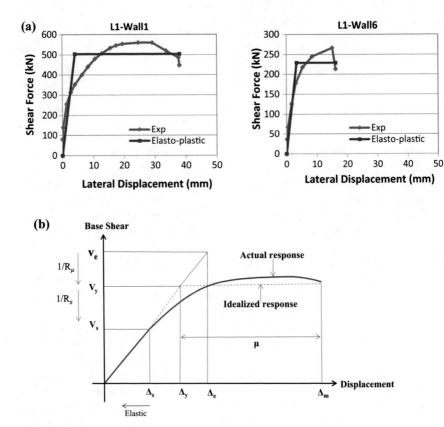

Fig. 3.3 Bi-linearization of capacity curve: **a** examples for the calculated data, **b** bilinear curve from ATC-40 (1996)

beam and the wall flexural compression zone, although it was designed and expected to fail in flexure. Its counterpart, wall L1-Wall5, however, subjected to f_m/f'_m of 0.10, displayed a ductile failure and rocking mechanism. Hence, the authors believe that until further research is available, it is recommended to keep f_m/f'_m smaller than 0.15.

Figure 3.5a, b present the effect of aspect ratio and PT bar spacing on the ductility of the considered PT-MWs. As shown, in ungrouted and partially grouted walls, as the aspect ratio increases and as the PT bar spacing decreases, the ductility increases. According to Fig. 3.5a, b for grouted walls with or without confinement or mild steel there is not any specific trend. As presented in Table 3.2, none of the ungrouted or partially grouted walls exhibited a flexural failure and they failed mainly in shear. According to the limited available data, it seems that the effect of aspect ratio and PT bar spacing on the ductility is more significant in ungrouted and partially grouted walls than with fully grouted walls.

Table 3.6 Ductility factor and fundamental period of walls

	Wall	Δ_y (mm)	Δ_{max} (mm)	μ	F_y (kN)	T (s)
Fully grouted	L1-Wall1	4.06	37.8	9.3	503.6	0.10
	L1-Wall2	0.94	20.78	22.2	429.9	0.10
	L1-Wall3	1.23	27.33	22.3	333.9	0.10
	L1-Wall4	1.03	35.53	34.5	394.8	0.10
	L1-Wall5	2.14	42.66	19.9	167.7	0.10
	L1-Wall6	3.34	16.01	4.8	228.8	0.10
	R-Wall1	3.22	97.01	30.1	311.5	0.10
Partially grouted	L1-Wall7	1.08	10.74	9.9	105.2	0.10
	W-Wall1	2.07	24.67	11.9	93.7	0.10
	W-Wall2	2.3	31.2	13.6	36.4	0.10
Un grouted	L1-Wall8	2.35	8.18	3.5	81.5	0.10
	R-Wall5	3.98	21.2	5.3	214.2	0.10
Bonded	R-Wall4	4.7	73.6	15.7	326.1	0.10
With confinement plates or supplemental mild steel	R-Wall2	2.28	224.3	98.4	324.7	0.10
	R-Wall3	4.95	127.95	25.8	346.3	0.10
	L2-Wall1	1.2	32.18	26.8	222.3	0.10
	L2-Wall2	2.35	30.87	13.1	369.3	0.10
	L2-Wall3	1.81	71.38	39.4	315.6	0.10
	L2-Wall4	19.38	343.65	17.7	393.6	0.10
	L2-Wall5	2.68	32.56	12.1	352.7	0.10
	L3-Wall1	5.72	82.32	14.4	196.5	0.17
	L3-Wall2	4.48	80.14	17.9	160.6	0.17
With opening	E-Wall1	2.79	52.53	18.8	180.6	0.09
	E-Wall2	1.52	40.78	26.8	140	0.09
	E-Wall3	12.3	42.1	3.4	191.7	0.09
	W2-WallD5	6.3	34.5	5.5	66.6	0.10
	W2-WallD6	7.3	28.2	3.9	61.3	0.10

3.6 Response Modification Factor

The response modification factor, R, is a constant accounting for the ductility and energy dissipation of the structure and is used to reduce the seismic load to a strength level design force. The International Building Code, IBC (2009) allows specially detailed structural components, which exhibit inelastic ductile behavior with high-energy dissipation, to undergo large deformations with high R values. The R-factor, is determined as follows (Uang 1991; Schmidt and Bartlett 2002; Asgarian and Shokrgozar 2009)

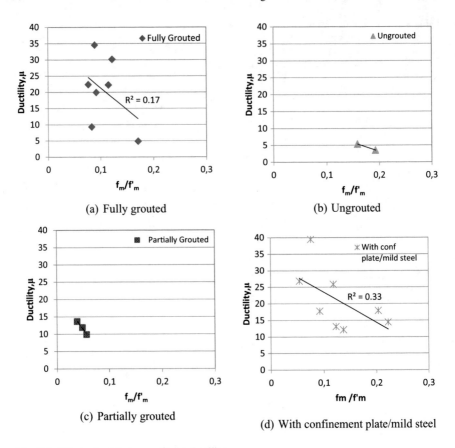

Fig. 3.4 Effect of axial stress ratio on ductility

$$R = R_s R_\mu R_R \tag{3.16}$$

where, R_R is a redundancy factor, R_μ is a reduction factor due to ductility and R_S is an over-strength factor.

The redundancy factor, R_R is introduced to account for the number and distribution of plastic hinges. R_μ can be determined either using nonlinear time history analyses or based on experimental data. The latter option is adopted in the current study. Figure 3.3b, presents a sample backbone and equivalent bilinear curve where the yield strength, yield displacement and maximum displacement are denoted by V_y, Δ_y and Δ_m, respectively. V_e corresponds to the elastic response strength of the system. As shown in the figure, the R_μ factor can be determined as V_e/V_y. Although the V_y value can be determined using bilinear idealized curves, the value of V_e is unknown. Over the last three decades, different equations have been developed to establish a relationship between R_μ and displacement ductility, μ, usually referred to as R–μ–T relationships, where T is the natural period of the structure (e.g.

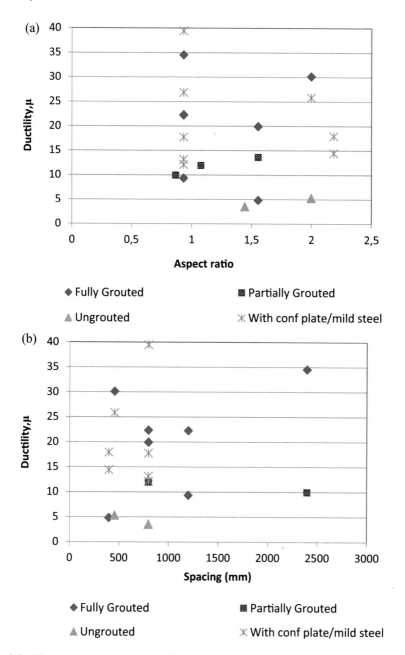

Fig. 3.5 Effect of **a** aspect ratio, and **b** PT bar spacing, on ductility

Newmark and Hall 1982; Riddell et al. 1989; Vidic et al. 1994; Nassar and Krawinkler 1991; Fajfar 2002). It was observed that in the medium and long period ranges, the equal displacement rule applies, i.e. displacement of the inelastic system and corresponding elastic system having the same period is almost the same (Newmark and Hall 1982). Hence, for medium to long period ranges, R_μ can be considered equal to the displacement ductility. For short period structures the ductility is higher than the R-factor. In Fajfar's expression, which is a modified version of an equation proposed by Vidic et al. (1994), the R–μ–T relationship was derived from a statistical study of a stiffness-degrading system with 10% strain hardening and 5% mass-proportional damping as mathematical models of damping. To calculate the reduction factor, a bilinear model and a stiffness degrading Q-model were analyzed. In the current study the following expressions, proposed by Fajfar (2002) were used to determine the R_μ-factor:

$$\begin{aligned} R_\mu &= (\mu - 1)\frac{T}{T_c} + 1 & T < T_c \\ R_\mu &= \mu & T \geq T_c \end{aligned} \tag{3.17}$$

where, T is the fundamental period of the structure and T_c is the characteristic period of the ground motion.

Other relationships for the R-factor have been proposed based on the equal energy concept. However, while numerous research and standards admit that R should be a function of T (Newmark and Hall 1982; Nassar and Krawinkler 1991; Vidic et al. 1994; Fajfar 2002; FEMA P695 2009), the current version of ASCE 7 (2010) utilizes period-independent R factors. FEMA P695 recommends directing future research to thoroughly investigating the question of period-dependent R factors and whether or not they should be considered for use in future versions of ASCE/SEI 7-10 (FEMA P695 2009). For short period ranges, the response factor is sensitive to the system period. As a consequence, a period-dependent seismic performance factor presents a more realistic response for structures with a shorter range of periods (FEMA P695 2009), and hence was selected in this study.

Figure 3.6 presents the R_μ values versus T, obtained from Eq. 3.17, for a range of displacement ductility values. All the curves converge to μ when the period of the system increases. As seen in the figure, for $T/T_c < 1$, the R_μ value increases almost linearly from 1.0 for a rigid system to approximately μ at $T/T_c = 1$. For longer period systems ($T/T_c > 1$), the R_μ values are less dependent on the system period, and can be considered to be constant.

T_c is defined as the transition period where the constant acceleration region of the response spectrum (the short-period range) progresses to the constant velocity region of the spectrum (the medium-period range) (Fajfar 2002). T_c is affected by the soil type, response acceleration parameters, seismicity, distance to faults, direction of faults and the location of the site. T_c is defined as S_{Ds}/S_{D1} where S_{Ds} and S_{D1} are design spectral response acceleration parameters and are defined by the following expressions (ASCE 7 2010):

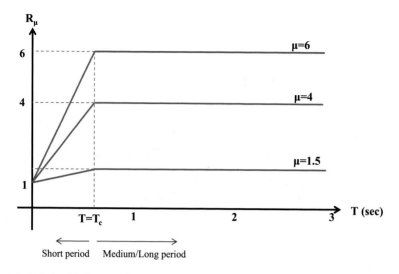

Fig. 3.6 Relationship between R_μ and T in Fajfar expression

$$S_{Ds} = \frac{2}{3}F_aS_s \tag{3.18}$$

$$S_{D1} = \frac{2}{3}F_vS_1 \tag{3.19}$$

where S_s and S_1 are earthquake spectral response accelerations at short periods and 1-second period, respectively. These values are given by the acceleration contour maps, for example the ones provided in ASCE 7 (2010). F_a and F_v are site amplification factors for different soil classes.

Post-tensioned masonry structures would have short-to-medium periods. Hence, T_c in Eq. 3.17 is an influential parameter. A statistical analysis was performed to evaluate the range of T_c for different site classes. The S_{1d}, S_{sd} and T_c values for 310 cities throughout the U.S. having different seismicity and soil classes taken from H-18-8 (2013) were calculated. The tables and calculations are provided in Appendix A. According to the results of the statistical analysis, the average T_c values were found to equal 0.35, 0.49, 0.53 and 0.57 for site classes A–B, C, D and E, respectively. The minimum values of T_c were found to equal 0.19, 0.27, 0.31 and 0.31 and maximum values were equal to 0.58, 0.82, 0.87 and 1.52 for the same order of site classes.

Figure 3.7 indicates the cumulative frequency and the distribution of the T_c values of the regions considered. From Fig. 3.7, the 95th percentile of T_c was determined to be equal to 0.46, 0.66, 0.69 and 1.01 for site classes A–B, C, D and E, respectively. According to Eq. 3.17, as T_c increases the R_μ-factor and hence the R-factor decreases. Considering the 95th percentile of T_c ensures that in more than

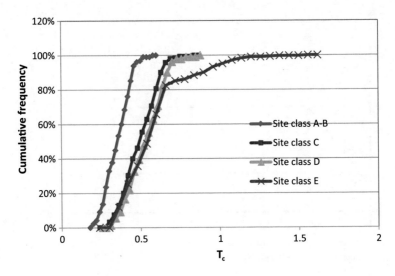

Fig. 3.7 Cumulative frequency versus T_c

95% of cases, the actual R-factor is higher than the values determined by the designer, which yields a safer design compared with the average value.

The cumulative curve in Fig. 3.7 indicates that there are just a few regions beyond the calculated 95th percentile, implying that the considered values yield a safe design. In this study the 95th percentile of T_c are considered in order to determine R-factors.

The fundamental period of a structure, T, in Eq. 3.3, can be estimated from Eq. 3.20 (ASCE 7 2010):

$$T = C_t h^{3/4} \tag{3.20}$$

C_t can be taken as 0.0488 for masonry structures. The height of the walls considered in the data base, h, ranged from 2.44 to 5.25 m. Table 3.6 presents the calculated period of the walls using the above expression. As shown, the period ranged between 0.09 and 0.17, corresponding to a short-medium period range. It is worth noting that these periods are typical for single to two story masonry houses.

3.7 Overstrength Factor (R_s)

There are several factors that contribute to the overstrength of the system and individual components, designated by the overstrength factor R_s. These include the ratio of actual material strength to the nominal strength, strain hardening of reinforcing steel, member selection (member oversize), minimum requirements by codes regarding proportioning and detailing, and the strain rate effect (Uang 1991).

In the current study, the size effect, material overstrength effect, strain hardening effect, and strain rate effect were considered through the F_1 to F_4 factors in Eq. 3.21:

$$R_s = F_1 F_2 F_3 F_4 \qquad (3.21)$$

The factor F_1 accounts for selecting PT bars and strands that are available in the market which often exceed the required steel area. F_1 has been assumed to be equal to 1.05 (Mitchell et al. 2003). Factor F_2 represents material over strength which is attributed to the difference between the nominal and actual material strengths. For steel bars, PT bars and strands, the material over strength factor can be different. For example, for ASTM a706 grade 60 rebar, the f_{py} can be anywhere between 413.7 MPa (60 ksi) and 537.8 MPa (78 ksi). Considering an average strength of 475.7 MPa (69 ksi), yields an over strength factor of about 1.15. The overstrength factor depends on the type of PT bars and strands. For example, the nominal and actual yield strength of Dywidag 23 mm-diameter reported by Laursen (2002) were 930 and 970 MPa, respectively corresponding to an overstrength factor of 1.04. However, the nominal and actual yield strength of 15.2 mm-diameter strands were 1520 and 1785 MPa respectively, corresponding to an overstrength factor of 1.09 (Laursen 2002). Conservatively and until more data become available, in this study the value of F_2 was assumed to be 1.05.

The strain hardening of PT bars and strands was reflected in Eq. 3.21 through the F_3 factor. The ultimate to yield strength ratio is about 1.15–1.2 in PT bars and strands. Based on Laursen's tests (2002), the strain hardening for 23 mm-diameter and 15.2 mm-diameter bars were 1.16 and 1.17, respectively. However, in the available database, rupture of PT bars or strands did not happen prior to the failure of the wall, as the wall needed to undergo a significant displacement before PT bar/strand rupture would occur. Hence, the F_3 value depends on how much strength is developed in the PT bars or strand in the strain hardening region, which is a function of the ductility and displacement capacity of the wall. To investigate the extent of strength gained due to strain hardening just before wall failure, the available recorded force in PT bars of the specimens in the database were collected and presented in Table 3.7. While for fully grouted walls, i.e., R-Wall1, R-Wall2 and R-Wall3, the PT bars yielded and strain hardening occurred, for the ungrouted wall i.e., R-Wall5, the PT bars did not yield. According to the table, the f_{ps}/f_{py} ranges from 1.00 to 1.07 with an average value of 1.04. For the ungrouted wall, R-Wall5, which exhibited a brittle behavior, no strain hardening should be considered as the PT bar did not yield. In this study, F_3 is considered to be equal to 1.04 for fully grouted and 1.00 for any other wall type. For fully grouted walls, the combined effect of F_2 and F_3 is hence equal to 1.092 which is less than 1.15, used by Wu (2008) to account for the prevention of end-plate failure.

The strain rate imposed on masonry, which is represented by the F_4 factor in Eq. 3.21, affects the material strength and behavior (Scott et al. 1982; Priestley and Elder 1983). This effect is higher in ductile elements that have excellent confinement of masonry (Asgarian and Shokrgozar 2009). Scott et al. (1982) conducted

Table 3.7 Calculating f_{ps}/f_{py} ratio

Wall	PT Bar No	f_{py} (MN)	f_{ps} (MN)	f_{ps}/f_{py}	$(f_{ps}/f_{py})_{avg}$
R-Wall1	PT bar 1	0.503	0.517	1.03	1.04
	PT bar 2	0.499	0.523	1.05	
	PT bar 3	0.508	0.517	1.02	
R-Wall2	PT bar 1	0.54	0.567	1.05	
	PT bar 2	0.506	0.506	1.00	
	PT bar 3	0.524	0.563	1.07	
R-Wall3	PT bar 1	0.478	0.508	1.06	
	PT bar 2	Did not yield ($f_s < f_y$)		1.00	
	PT bar 3	0.489	0.51	1.04	
R-Wall4	PT bar	Did not yield ($f_s < f_y$)		1.00	1.00

tests on confined and unconfined reinforced concrete units at low and high strain rate and found that the strength increased by 25% with a high strain rate of 0.0167/s compared to a slower strain rate (Priestley and Elder 1983). This indicates that the strain rate is very significant in estimating the strength of masonry. However this rapid strain rate is three times higher than the adopted strain-rate in masonry standards (Priestley and Elder 1983). Reducing the strain rate by a factor of 10, increased the strength by 13% (Scott et al. 1982). Priestley and Elder (1983) observed a strength increase of 20.5% for a strain rate of 0.005–0.006/s compared with a slow strain rate of 0.000005/s. Priestley and Elder compared their results with those reported by Scott et al. (1982), and considering the standard slow strain rate, they recommended 17% strength enhancement to account for the high strain rate. This value is adopted in this study as a strength enhancement factor due to strain rate and hence $F_4 = 1.17$. Considering the above mentioned factors, R_s becomes 1.34 for fully grouted and 1.23 for other wall types.

3.8 Seismic Response Factors of PT-MWs

Tables 3.8, 3.9, 3.10 and 3.11 present the R_μ and R factors calculated for the database walls using Fajfar's (2002) expression. The average values of R_μ and R are also presented in Tables 3.8, 3.9, 3.10 and 3.11. According to Table 3.8, for fully grouted walls, the average R-factor for different site classes ranged from 4.3 to 7.8. According to Tables 3.9, 3.10 and 3.11, depending on the site class, the average R-factor ranged from 1.63 to 2.11 for ungrouted walls, 2.55 to 4.12 for partially grouted walls, 4.31 to 7.87 for walls with confinement plate/supplemental mild steel, and 2.63 to 4.18 for walls with openings.

Figure 3.8 presents the calculated R-factors versus T_c values for the 31 wall specimens considered. As shown in the figure, R factors are functions of T_c. As the considered walls have a short-to-intermediate period range, the R-factor is sensitive

Table 3.8 R_μ and R-factor—fully grouted walls

	R_μ				R-factor			
	$T_c = 0.46$ s	$T_c = 0.66$ s	$T_c = 0.69$ s	$T_c = 1.01$ s	$T_c = 0.46$ s	$T_c = 0.66$ s	$T_c = 0.69$ s	$T_c = 1.01$ s
L1-Wall1	2.80	2.26	2.20	1.82	3.76	3.03	2.95	2.44
L1-Wall2	5.61	4.21	4.07	3.10	7.52	5.64	5.46	4.15
L1-Wall3	5.63	4.23	4.09	3.11	7.54	5.66	5.48	4.17
L1-Wall4	8.28	6.08	5.86	4.32	11.10	8.14	7.85	5.78
L1-Wall5	5.11	3.86	3.74	2.87	6.85	5.18	5.01	3.85
L1-Wall6[a]	1.83	1.58	1.55	1.38	2.25	1.94	1.91	1.69
R-Wall1	7.33	5.41	5.22	3.88	9.82	7.25	6.99	5.20
Average	5.79	4.34	4.20	3.18	7.76	5.82	5.62	4.27

[a]Wall L1-Wall6 displayed a shear failure mode and hence is not considered in calculation of average values for R and R_μ

Table 3.9 R_μ and R-factor—partially grouted and ungrouted walls

		R_μ				R-factor			
		$T_c = 0.46$ s	$T_c = 0.66$ s	$T_c = 0.69$ s	$T_c = 1.01$ s	$T_c = 0.46$ s	$T_c = 0.66$ s	$T_c = 0.69$ s	$T_c = 1.01$ s
Ungrouted	L1-Wall8	1.54	1.38	1.36	1.25	1.89	1.69	1.67	1.53
	R-Wall5	1.90	1.62	1.60	1.41	2.33	2.00	1.96	1.73
	Average	1.72	1.50	1.48	1.33	2.11	1.85	1.82	1.63
Partially grouted	L1-Wall7	2.94	2.35	2.30	1.88	3.62	2.90	2.82	2.32
	W-Wall1	3.37	2.65	2.58	2.08	4.15	3.26	3.17	2.56
	W-Wall2	3.73	2.90	2.82	2.24	4.59	3.57	3.47	2.76
	Average	3.35	2.64	2.57	2.07	4.12	3.24	3.16	2.55

Table 3.10 R_μ and R-factor -walls with confinement plate or supplemental mild steel

	R_μ				R-factor			
	$T_c = 0.46$ s	$T_c = 0.66$ s	$T_c = 0.69$ s	$T_c = 1.01$ s	$T_c = 0.46$ s	$T_c = 0.66$ s	$T_c = 0.69$ s	$T_c = 1.01$ s
R-Wall2[a]	21.17	15.06	14.45	10.19	28.37	20.18	19.36	13.65
R-Wall3	6.15	4.59	4.43	3.34	8.24	6.15	5.94	4.48
L2-Wall1	6.61	4.91	4.74	3.55	8.85	6.58	6.35	4.76
L2-Wall2	3.64	2.84	2.76	2.20	4.87	3.80	3.69	2.95
L2-Wall3	9.35	6.82	6.57	4.80	12.53	9.14	8.80	6.44
L2-Wall4	4.63	3.53	3.42	2.66	6.21	4.73	4.59	3.56
L2-Wall5	3.42	2.69	2.61	2.10	4.59	3.60	3.50	2.82
L3-Wall1	5.95	4.45	4.30	3.26	7.98	5.97	5.76	4.36
L3-Wall2	7.25	5.35	5.16	3.84	9.71	7.17	6.92	5.15
Average[a]	5.87	4.40	4.25	3.22	7.87	5.89	5.69	4.31

[a]Wall R-Wall2 is not considered in calculation of average values for R and R_μ, as the result was comparatively much higher than other test results

Table 3.11 R_μ and R-factor—walls with openings

	R_μ					R-factor				
	$T_c = 0.46$ s	$T_c = 0.66$ s	$T_c = 0.69$ s	$T_c = 1.01$ s	$T_c = 0.46$ s	$T_c = 0.66$ s	$T_c = 0.69$ s	$T_c = 1.01$ s
E-Wall1	4.52	3.45	3.34	2.60	6.05	4.62	4.48	3.49
E-Wall2	6.09	4.55	4.40	3.32	8.16	6.10	5.89	4.45
E-Wall3	1.48	1.33	1.32	1.22	1.98	1.79	1.77	1.63
W2-WallD5	1.93	1.65	1.62	1.42	2.58	2.21	2.17	1.91
W2-WallD6	1.59	1.41	1.40	1.27	2.13	1.89	1.87	1.70
Average	3.12	2.48	2.41	1.97	4.18	3.32	3.23	2.63

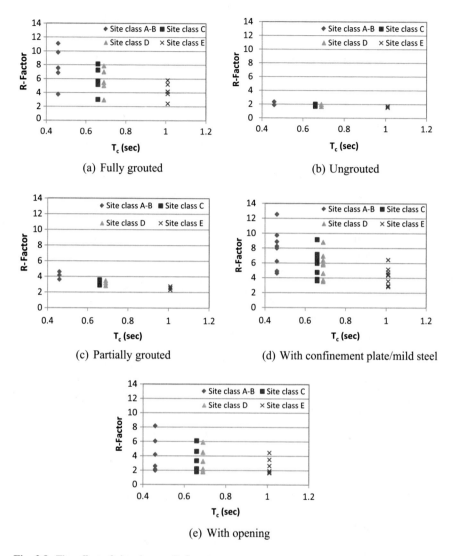

Fig. 3.8 The effect of site class on R-factor

to the sub-soil type. As a general trend for all walls, as T_c increases and as the sub soil changes from bedrock to soft soil, the R-factor decreases. This trend is more significant in fully grouted walls and walls with confinement plate/supplemental mild steel. Figure 3.9 compares the average R-factor for different wall types which are also presented in Tables 3.8, 3.9, 3.10 and 3.11. As shown in the figure, the variations of R-factor for each wall type are significant. As an example and for $T_c = 0.46$ s, the R-factor ranged from 2.25 to 11.1, 1.89 to 2.33, 3.62 to 4.59, 4.59 to 28.37, 1.98 to 8.16 for fully grouted walls, ungrouted walls, partially grouted walls,

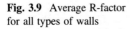

Fig. 3.9 Average R-factor
for all types of walls

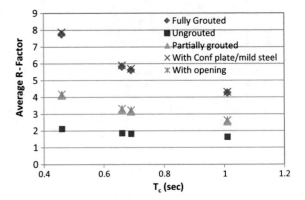

walls with confinement plates or mild steel, and walls with openings, respectively. Hence, within each wall type the variations in the R-factor are significant since the R-factor does not depend solely on the wall type but also on the applied level of post-tensioning in the bars. For example, the calculated average R-factor of the walls with confinement plate/mild steel is slightly higher than fully grouted walls which may be counter-intuitive. However, many of these specimens had high initial post-tensioning forces in PT bars, leading to early yielding of the post-tensioning bars, which resulted in a considerable ductility and overshadowed the ductility increase due to confinement plates and/or mild steel. Consequently, it is suggested that the same value of R-factor should be considered for fully grouted walls with or without confinement plate/mild steel. It should be noted that the available number of test specimens in each group of wall types is limited and hence these results need to be reviewed when more data become available.

All of the R-factors determined and presented so far are determined based on the 95th percentile of T_c. If the maximum value of T_c is considered in Eq. 3.17 and following the same procedure presented before, the average R-factor was found to range from 3.46 to 6.77 for six fully grouted walls, 1.50 to 1.93 for two ungrouted walls, 2.10 to 3.52 for three partially grouted walls, 3.49 to 6.86 for eight walls with confinement plate/supplemental mild steel, and 2.31 to 3.78 for five walls with openings depending on soil type. Until further data is available, it is recommended to assign minimum R-factors of 3.5 for fully grouted (with/without confinement plate/ supplemental mild steel), 2.0 for partially grouted and 1.5 for ungrouted PT masonry walls, respectively.

3.8.1   Effect of Supplemental Mild Steel

Walls R-Wall3 and R-Wall1 were identical; both walls had an aspect ratio of two and were subjected to an axial stress of 2.69 MPa and the bars had an initial post-tensioning stress of 70% of the yield strength. However, within the database,

R-Wall3 is the only wall including supplemental mild steel. A mild steel ratio of 0.139% was provided through 4#4 (12 mm) grade 60 bars which were placed near the wall toe and heel. These bars extended 914 mm into the footing and wall with 609 mm as the unbonded length (Rosenboom 2002).

From Tables 3.8 and 3.10, while the R-factor calculated for R-Wall1 ranged from 5.20 to 9.82 depending on the T_c value, adding mild steel in the wall R-Wall3 reduced the R value range to 4.48–8.24 depending on the T_c value, a drop of 16–19%. Comparing the strength and ductility of these two walls through Tables 3.2 and 3.6, also reveals that while supplemental mild steel increased the strength by 10%, it reduced the ductility by about 15%. Adding mild steel increased the yield displacement by approximately 53% compared to wall R-Wall1 without mild steel. However, the mild steel increased the ultimate displacement by only 31% compared to the specimen without mild steel. This led to a reduction in the displacement ductility in the case of the specimen having mild steel compared to the one without mild steel. It is worth noting that the post-tensioning bars in both specimens yielded early at drifts of 2 and 1% for specimens with and without mild steel, respectively. Yielding of post-tensioning led to losses of post-tensioning force and hence the effect of mild steel was not pronounced.

3.8.2   Effect of Grouting

According to Figs. 3.8 and 3.9, the ungrouted walls exhibited the lowest R-factor as expected. Moreover, the R-factor is less dependent on soil type and remained around 2.0, with a lesser variation compared to other walls. Ungrouted post-tensioned walls displayed a brittle behavior, characterized by a relatively small R-factor and ductility and can be considered as ordinary plain masonry shear walls in terms of seismic design factors.

The R-factors determined for partially grouted walls and fully grouted walls with openings are similar and both are much less than that of fully grouted walls. It seems that the same R-factors for partially grouted walls can be applied to fully grouted walls with openings. As predicted and as presented in Fig. 3.9, the average response modification factor for partially grouted walls is somewhere between fully grouted and ungrouted walls. In partially grouted walls it is important to determine how many of the cells are grouted. As the number of grouted cells increases from zero to fully grouted, the R-factor increases from a small value corresponding to an ungrouted masonry wall to a high value corresponding to a fully grouted masonry wall. To investigate if the R-factor of partially grouted walls can be estimated using ungrouted and fully grouted wall R-factors, the calculated values from the tests and the values obtained from linear interpolation (weighted average) are presented in Table 3.12. According to the table, the R-factor estimated from linear interpolation falls within ±20% of the calculated values obtained from test results. Hence, the R-factor for fully grouted and ungrouted walls can be used to determine the R-factors of partially grouted walls, by interpolating between fully grouted and ungrouted R-factors, depending on the percentage of the grouted cells.

Table 3.12 Linear interpolation of **R**-factor for partially grouted walls

Partially grouted walls	Portion of grouted cavities	Calculated from test, R				Linear interpolation, R_{int}				R/R_{int}			
		$T_c = 0.46$ s	$T_c = 0.66$ s	$T_c = 0.69$ s	$T_c = 1.01$ s	$T_c = 0.46$ s	$T_c = 0.66$ s	$T_c = 0.69$ s	$T_c = 1.01$ s	$T_c = 0.46$ s	$T_c = 0.66$ s	$T_c = 0.69$ s	$T_c = 1.01$ s
L1-Wall7	2/15	2.87	2.37	2.33	1.98	2.94	2.35	2.30	1.88	0.97	1.01	1.01	1.05
W-Wall1	1/3	3.85	3.07	2.99	2.44	3.37	2.65	2.58	2.08	1.14	1.16	1.16	1.17
W-Wall2	4/13	4.00	3.17	3.09	2.51	3.73	2.90	2.82	2.24	1.07	1.09	1.09	1.12

3.8.3   Effect of Initial to Yield Stress Ratio in the Bar

According to Tables 3.6, 3.8, 3.9, 3.10 and 3.11 different ranges of ductility and R-factors are obtained for the walls within the database. Various factors contribute to the ductility of the specimens within a group. The initial to yield stress ratio of PT bars, f_i/f_{py}, seems to have a significant influence on the ductility of the PT-MWs. For example, walls L1-Wall2 and L1-Wall4 have the same characteristics but wall L1-Wall2 included three PT bars while wall L1-Wall4 had two PT bars. The size of the PT bars was the same in both tests. In order to apply the same level of axial stress to wall L1-Wall2 and L1-Wall4, an initial post-tensioning force of 555 and 757 kN was applied to the bars, corresponding to an initial to yield stress ratio of 0.57 and 0.78, respectively. As shown in Tables 3.2 and 3.6, while wall L1-Wall2 exhibited a higher strength, it presented a less ductile behavior; which can be attributed to the ratio of f_i/f_{py}. Larger values of f_i/f_{py} result in yielding of the PT bars at relatively small drifts compared to those of similar walls having the same post-tensioning force but with smaller f_i/f_{py} values. A lesser value of f_i/f_{py} provides more capacity for the PT bars to elongate prior to yielding. Consequently, starting with the same initial total post-tensioning force at the beginning of the test, the specimens with smaller f_i/f_{py} develop more axial load during the test. For example, the ultimate total post-tensioning force in wall L1-Wall2 and L1-Wall4 were 800 and 650 kN corresponding to increases of 16 and 3.5% of the total initial PT force, respectively. Although a higher PT force results in a higher strength level, it causes the strength to degrade rapidly following the peak. This is mainly because of the adverse effect of axial stress on the ductility as shown in Fig. 3.4. A higher rate of strength degradation, results in smaller values of maximum displacement Δ_m. To investigate the general trend and effect of f_i/f_{py} based on the available database, Fig. 3.10a, b shows the values of f_i/f_{py} versus ductility and R-factor, respectively. While there is a general trend of the data that supports the abovementioned conclusion obtained from walls L1-Wall2 and L1-Wall4, there is still a strong scatter in the data and more investigation is required.

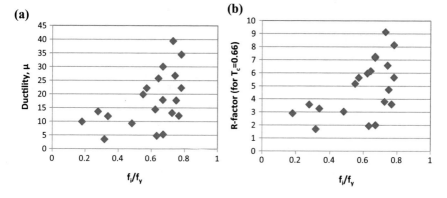

Fig. 3.10 Effect of f_i/f_y on **a** ductility, and **b** R-factor

3.8.4 Effect of Confinement Plates

The effect of confinement plates on the R value can be better understood by comparing the response of walls L1-Wall3 and R-Wall1 with their counterpart walls L2-Wall2 and R-Wall2 which include confinement plates, respectively. As shown in Table 3.2, the strength of the L2-Wall2 and R-Wall2 increased slightly (by about 5–6%), compared with their unconfined counterparts. However, the displacement response is not consistent. While in wall R-Wall2 the ductility increased compared with the unconfined counterpart, R-Wall1, in wall L2-Wall2 the ductility reduced compared to wall L1-Wall3. Based on the limited available data and until further research is available, it is recommended to ignore the effect of a confinement plate on increasing ductility, and hence apply the same R-factor to the confined walls as the walls without confinement plates.

ASCE 7 (2010) considers an R-factor of 1.5 for all types of PT-MWs. Comparing this value with the R-factors obtained in this study reveals that, although a value of 1.5 is reasonable for ungrouted PT-MWs, it is a conservative value for partially grouted and fully grouted PT-MWs. R-factors of 2.5 and 3 are recommended for partially grouted and fully grouted PT-MWs, respectively.

3.9 Displacement Amplification Factor

Displacements from elastic analysis obtained from reduced forces are amplified by the displacement amplification factor c_d to account for inelastic effects. Equating the deflection amplification factor, c_d, to the R factor is based on the equal displacement assumption (Newmark and Hall 1982). This is consistent with research findings for systems with nominal (5% of critical) damping and fundamental periods greater than T_c (FEMA P695 2009). While the equal displacement assumption provides a good estimate for prediction of the c_d factor in structures having longer natural periods (Watanabe and Kawashima 2004), it is recognized that for short-period systems ($T < T_c$) inelastic displacement generally exceeds elastic displacement. In these structures, such as single storey buildings, the average values of displacement amplification factors sharply increase as the natural periods decrease (Uang and Maarouf 1994; Watanabe and Kawashima 2004; FEMA P695 2009) and equal energy assumptions provide more realistic results compared to equal displacement assumptions. Considering equal energy assumptions, the c_d factor can be determined from the following equation (Uang and Maarouf 1994; Watanabe and Kawashima 2004):

$$c_d = \frac{\mu}{\sqrt{2\mu - 1}} \tag{3.22}$$

The c_d values are presented in Table 3.13 obtained using Eq. 3.22. It can be seen that while for fully grouted walls the c_d-factor ranges between 1.9 and 4.8, with an average of 3.5, the average c_d factor for ungrouted and partially grouted walls is 1.8 and 2.9, respectively. The average c_d value is determined to equal 4.2 and 2.6 for supplemental mild steel/confinement and walls with openings, respectively.

ASCE 7 (2010) considers $c_d = 1.75$ for all types of post-tensioned masonry walls. Comparing this value with the c_d-factors determined in this study, reveals that although a value of 1.75 is reasonable for ungrouted post-tensioned walls, it provides a conservative and very conservative prediction of c_d-factor for partially and fully grouted walls, respectively.

Table 3.13 Displacement modification factor

	Wall	$c_d = \frac{\mu}{\sqrt{2\mu-1}}$	Average
Fully grouted	L1-Wall1	2.6	3.6
	L1-Wall2	3.9	
	L1-Wall3	3.9	
	L1-Wall4	4.8	
	L1-Wall5	3.7	
	L1-Wall6	1.9	
	R-Wall1	4.5	
Partially grouted	L1-Wall7	2.6	2.9
	W-Wall1	2.9	
	W-Wall2	3.1	
Un grouted	L1-Wall8	1.6	1.8
	R-Wall5	2	
Bonded	R-Wall4	2.85	2.85
With confinement plates or supplemental mild steel	R-Wall2	8.1	4.2
	R-Wall3	4.2	
	L2-Wall1	4.3	
	L2-Wall2	3	
	L2-Wall3	5.2	
	L2-Wall4	3.5	
	L2-Wall5	2.9	
	L3-Wall1	3.2	
	L3-Wall2	3.5	
With opening	E-Wall1	3.6	2.6
	E-Wall2	4.3	
	E-Wall3	1.6	
	W2-WallD5	2	
	W2-WallD6	1.7	

3.10 Conclusions

This study has examined the behavior of PT-MWs according to the test results of 31 tested wall specimens. The accuracy of ignoring elongation of PT bars in the MSJC (2013) standard in predicting the strength of these walls was investigated based on the available test results. Moreover, the structural response parameters including ductility, response modification factor and displacement amplification factor have been studied. Based on the results the following conclusions are reached:

- The flexural strength of fully grouted unbonded PT-MWs is under predicted by MSJC (2013) due to ignoring the elongation of PT bars in the flexural strength expression of MSJC (2013). The under prediction ranged from 58 to 99%, with an average of 80%.
- Based on the shear expression provided in MSJC (2013), the shear strength of partially grouted and ungrouted post-tensioned walls was over predicted by 12–86%. Hence, a revised shear strength equation is urgently needed.
- According to the test results of the single bonded specimen of the database, using the strain compatibility method to estimate the strength of bonded post-tensioned walls resulted in an acceptable prediction.
- The axial stress ratio has a prominent effect on the ductility. Based on the limited available data, to provide a ductile response, it is recommended to limit the axial stress ratio to a value of 0.15.
- The response modification factor of post-tensioned walls is a function of site class. The average R-factors ranged between 4.27–7.76, 1.63–2.11, 2.55–4.12, 4.31–7.87 and 2.63–4.18 for fully grouted walls, ungrouted walls, partially grouted walls, walls with confinement plate/supplemental mild steel and walls with openings, respectively.
- Relatively high values of R-factors were obtained for fully and partially grouted walls (with or without confinement plate or supplemental mild steel). It is recommended that R-factors of 2.5 and 3.0 for partially grouted and fully grouted PT-MWs, respectively, should be used.
- Ungrouted pre-stressed walls displayed brittle behavior, characterized by a relatively small R-factor and ductility. These walls exhibited a limited displacement capacity and can be considered as ordinary plain masonry shear walls with minimal ductility. The recommended R-factor for these walls is 1.5.
- The ductility and response modification factor for walls with openings is much less than that of fully grouted walls.
- Bonded post-tensioned walls exhibited lower ductility and R-factors compared with their unbonded counterparts.

- For fully grouted walls the c_d-factor was determined to be equal to 3.5. The average values for ungrouted, partially grouted, supplemental mild steel/ confinement and walls with openings were found to be equal to 1.8, 2.9, 4.5 and 2.6, respectively.
- In almost all of the post-tensioned masonry walls tested so far that failed due to flexure, the PT bars have yielded. This results in an increased ductility, energy dissipation and a higher response modification factor. For these walls the supplemental mild steel is not required, as it does not increase the ductility of the system. It seems that the effect of supplemental mild steel on ductility is negligible, unless the PT bar does not yield before the failure of the wall.

While the research presented in this manuscript presents the first systematic approach to determine the seismic parameters for post-tensioned masonry walls (PT-MWs) and determine the influence of different parameters on the seismic response of PT-MWs, there is an urgent need to enlarge the number of specimens in the database to confirm the conclusions from the current study. Moreover, this manuscript focused on the component level and hence more research is required focusing on the response at the system level.

References

ASCE (2010) Minimum design loads for buildings and other structures. Standard ASCE/SEI 7-10, American Society of Civil Engineers, Reston, VA

Asgarian B, Shokrgozar HR (2009) BRBF response modification factor. J Constr Steel Res 65 (2):290–298

ATC-40 (1996) Seismic evaluation and retrofit of concrete buildings. Applied Technology Council, report ATC-40. Redwood City, 8–31

Bean J (2007) Mechanics and behavior of slender, post-tensioned masonry walls to transverse loading. Ph.D. dissertation, University of Minnesota, Minnesota, MN, USA

ElGawady MA, Sha'lan A (2011) Seismic behavior of self-centering precast segmental bridge bents. J Bridge Eng ASCE 16(3):328–339

ElGawady MA, Booker AJ, Dawood H (2010) Seismic behavior of post-tensioned concrete-filled fiber tubes. J Compos Const, ASCE 14(5):616–628

Erkmen B, Schultz AE (2009) Self-centering behavior of unbonded precast concrete shear walls. J Earthq Eng 13(7):1047–1064

Ewing B (2008) Performance of post-tensioned clay brick masonry walls with openings. Ph.D. thesis, North Carolina State University, Raleigh, NC, USA

Fajfar P (2002) Structural analysis in earthquake engineering—a breakthrough of simplified non-linear methods. In: 12th European conference on earthquake engineering, Elsevier, London, UK, Paper reference 843, Elsevier

FEMA P695 (2009) Quantification of building seismic performance factors. Federal Emergency Management Agency, Washington, DC

H-18-8 (2013) VA seismic design requirement, U.S. department of Veterans Affairs, Office of construction and facilities management

IBC (2009) International building code. International Code Council, Inc. (formerly BOCA, ICBO and SBCCI) vol 4051, pp 60478–65795

Laursen PPT (2002) Seismic analysis and design of post-tensioned concrete masonry walls. Ph.D. dissertation, Department of Civil and Environmental Engineering, University of Auckland, Auckland, New Zealand

Lissel SL, Shrive NG (2003) Construction of diaphragm walls post-tensioned with carbon fiber reinforced polymer tendons. In: Proceedings of the 9th North American Masonry conference (9NAMC), Clemson, SC, USA, pp 192–203

Masonry Standards Joint Committee (MSJC) (2013) Building code requirements for masonry structures, ACI 530/ASCE 5, TMS 402. American Concrete Institute, Detroit

Mitchell D, Tremblay R, Karacabeyli E, Paultre P, Saatcioglu M, Anderson DL (2003) Seismic force modification factors for the proposed 2005 edition of the National Building Code of Canada. Can J Civil Eng 30(2):308–327

Nassar AA, Krawinkler H (1991) Seismic demands for SDOF and MDOF systems. John A. Blume Earthquake Engineering Center, Department of Civil Engineering, Stanford University, California, USA

Newmark NM, Hall WJ (1982) Earthquake spectra and design. Technical Report, Earthquake Engineering Research Institute, Berkeley, California

Page A, Huizer A (1988) Racking behavior of pre-stressed and reinforced hollow masonry walls. Masonry Int 2(3):97–102

Priestley M, Elder D (1983) Stress-strain curves for unconfined and confined concrete masonry. ACI J Proc 80(3):192–201

Riddell R, Hidalgo P, Cruz E (1989) Response modification factors for earthquake resistant design of short period buildings. Earthq Spectra 5(3):571–590

Rosenboom OA (2002) Post-tensioned clay brick masonry walls for modular housing in seismic regions. M.S. thesis, North Carolina State University, Raleigh, NC, USA

Rosenboom OA, Kowalsky MJ (2004) Reversed in-plane cyclic behavior of post-tensioned clay brick masonry walls. J Struct Eng 130(5):787–798

Ryu D, Wijeyewickrema A, ElGawady M, Madurapperuma MAKM (2014) Effects of tendon spacing on in-plane behavior of post-tensioned masonry walls. J Struct Eng 140(4), CID:04013096

Schmidt B, Bartlett F (2002) Review of resistance factor for steel: data collection. Can J Civ Eng 29(1):98–108

Schultz AE, Scolforo MJ (1991) An overview of pre-stressed masonry. TMS J Masonry Soc 10 (1):6–21

Scott BD, Park R, Priestley MJN (1982) Stress-strain behavior of concrete confined by overlapping hoops at low and high strain rates. ACI J 79(1):13–27

Shrive NG (1988) Post-tensioned masonry-status & prospects. In: The Canadian society for civil engineering—annual conference, Calgary, Canada, pp 679–606

Uang CM (1991) Establishing R (or R_w) and c_d factors for building seismic provisions. J Struct Eng 117(1):19–28

Uang C, Maarouf A (1994) Deflection amplification factor for seismic design provisions. J Struct Eng 120(8):2423–2436

Vidic T, Fajfar P, Fischinger M (1994) Consistent inelastic design spectra: strength and displacement. Earthq Eng Struct Dynam 23(5):507–521

Watanabe G, Kawashima K (2004) An evaluation of the displacement amplification factors for seismic design of bridges. In: First international conference on urban Earthquake engineering, Center for Urban Earthquake Engineering, Tokyo Institute of Technology, Tokyo, Japan

Wight GD (2006) Seismic performance of a post-tensioned concrete masonry wall system. Ph.D. dissertation, Department of Civil and Environmental Engineering, University of Auckland, Auckland, New Zealand

Wight GD, Ingham JM (2008) Tendon stress in unbonded post-tensioned masonry walls at nominal in-plane strength. J Struct Eng 134(6):938–946

Wu Y (2008) Development of precast concrete and steel hybrid special moment-resisting frames. Ph.D. thesis, University of Southern California, Los Angeles, CA, USA

Chapter 4
Effect of Dimensions on the Compressive Strength of Concrete Masonry Prisms

This chapter investigates the accuracy of the height-to-thickness ratio (h/t) correction factors presented in the ASTM standard (ASTM C1314-03) and in other international standards using numerical finite element analysis. The FEM is calibrated with experimental results, and then a parametric study is performed to examine the effect of size on the strength of masonry prisms. Calibration of masonry material provided in this chapter is then used in developing finite element models of PT-MWs presented in Chap. 5.

4.1 Introduction

The compressive strength is the most significant property by which the quality of masonry is evaluated. Similar to concrete, the 28-day compressive strength is commonly considered as a standard strength by masonry codes worldwide. For masonry this value is determined by compression tests on standard prisms. To characterize the properties of a masonry member, the prism specimens must consist of more than one course of masonry unit and a layer of mortar. The same type of grout, mortar and block used in the construction of the masonry member must be used in the prism construction to appropriately represent the behavior.

The compressive strength of masonry is obtained using a compressive machine. During testing a level of friction is induced by the machine platen to the top and bottom surfaces of the prism as it provides restraint and prevents the expansion of the specimen, and hence, generates confinement around the two ends of the specimen. For a fully grouted prism with a small height-to-thickness (h/t) ratio, the

A modified version of this chapter has been published in the Journal of Advances in Civil Engineering Materials: Hassanli R., ElGawady M. A. and Mills J. E., Effect of dimensions on the compressive strength of concrete masonry prisms, Advances in Civil Engineering Materials, ASTM, 4(1), 175–201, 2015.

confining stresses are developed throughout the specimen height, hence influencing the compressive strength. If the lateral constraint is eliminated, the lateral confining force disappears and a splitting failure mode will govern the response (Malvar et al. 2004). While ideally a frictionless testing machine yields a more realistic compressive strength, in practice it is difficult to eliminate the frictional force. To obtain an accurate unconfined strength of masonry regardless of the effect of the testing machine and to account for the influence of varying sample dimensions, "strength correction factors" are introduced by masonry codes. These factors are h/t dependent parameters and convert the compressive strength of the tested prism to a standard size prism. Table 4.1 present the correction factors in some selected standards, namely (ASTM C1314-03b 2003): standard of USA; AS 4456.4: (2003) standard of Australia and CSA S304.1 (2004): standard of Canada. According to ASTM C1314-03b (2003), the h/t ratio of clay and concrete masonry prisms must be within the range of 2.0–5.0 and 1.33–5.0, respectively. From Table 4.1, it can be seen that in international codes in general an h/t ratio of 5.0 is considered as a reference ratio corresponding to a correction factor of 1.0. For clay masonry the ASTM C1314 considers an h/t ratio of 5.0 as reference, however, for concrete masonry, ASTM C1314 is the only standard which considers an h/t ratio of 2.0 as reference.

Using the same underlying principle for the correction factors, the correction factors for concrete masonry in the ASTM C1314 (2003) standard can be normalized for $h/t = 5.0$, to match with the other codes. Figure 4.1 compares the general relationship between h/t and corrections factors in different standards. As can be seen in the figure, in all standards, as the h/t ratio increases the correction factor increases.

Normalization of the concrete masonry prism in the ATSM standard based on $h/t = 2.0$ is mainly due to the following three reasons. Firstly, the clear height of cylinder compression machines that are typically used to test masonry prisms is appropriate is limited. Most machines accommodate prisms with $h/t = 2$, but not any taller than that.

Secondly, for concrete materials, an aspect ratio of 2.0 is perceived globally as a reference ratio. ASTM C1314 sets the same h/t ratio for concrete masonry as both concrete masonry and concrete are cement-based materials with fundamental

Table 4.1 Height-to-thickness, h/t, correction factors

ASTM C1314 concrete Masonry		1.33	1.5	2	3.0	4.0	5.0
		0.75	0.86	1.00	1.07	1.15	1.22

ASTM C1314 Clay Masonry	2.0	2.5	3.0	3.5	4.0	4.5	5.0
	0.82	0.85	0.88	0.91	0.94	0.97	1.00

CSA S304.1	2.0	3.0	4.0	5–10
	0.8	.090	0.95	1.00

AS4456.4	1.0	5.0
	0.7	1.0

Fig. 4.1 Height-to-thickness ratio correction factors

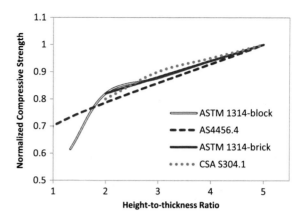

similarities in behavior and material properties. Thirdly, a lighter prism is easier to use for testing. In the US, if grouted prisms are to be tested, typically half-length units are used to minimize the weight of the prism. A two-course high prism with one bed joint is the minimum required to construct a prism for testing, giving a height-to-thickness ratio of more than two for most of the common size concrete units.

Figure 4.2a shows the development of stresses due to lateral confinement in a typical concrete cylinder specimen and a fully grouted masonry prism with a height-to-thickness ratio of 2.0. It can be seen that assuming an internal angle of friction, φ, of 60°, in a cylindrical specimen with h/t of higher than 2.0, there is an unaffected lateral stress zone in the central region of the specimen. This implies that the effect of top and bottom confinement due to friction with the machine platen is not contributing to the concrete strength for cylinder specimens with an aspect ratio of higher than 2.0. This is the reason why concrete codes, consider an aspect ratio of 2.0 as a reference and assign a correction factor of 1.0 to this aspect ratio, and why an aspect ratio of 2.0 is accepted globally as a standard for different sizes of concrete cylinders, e.g. cylinder with diameter × height of 50 mm × 100 mm, 100 mm × 200 mm, 150 mm × 300 mm.

According to ASTM C 42-90 (1992), in order to prevent the over-prediction of the actual unconfined strength of the concrete cylinders having aspect ratios of less than 2.0, the compressive strength obtained during testing must be multiplied by a strength reduction factor presented in Table 4.2.

This modification factor converts the compressive strength of a concrete cylinder with $h/t < 2$, to the strength of a standard size cylinder.

While in concrete cylinders having an aspect ratio of greater than 2.0, the confining effect is not affecting the mid-height of the specimen, the mechanism of stress developed in a masonry prism is challenging, mainly due to difference in cross sectional dimensions (Fig. 4.2b). Based on the available common concrete block sizes, it seems that considering an h/t of 2.0 as a reference in ASTM C1314 provides a degree of overestimation of the compressive strength.

Fig. 4.2 Effect of end platen on **a** concrete cylinder, and **b** masonry prism

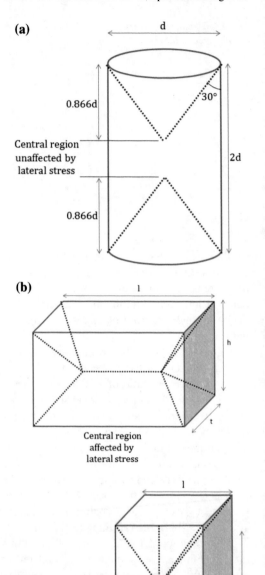

Table 4.2 Correction factor
for concrete cylinders

h/d	1.00	1.25	1.5	1.75	2.00
Strength correction factor	0.87	0.93	0.96	0.98	1.00

d diameter of the cylinder

Constructing a concrete test cylinder is practical by using standard sizes of cylindrical moulds; however, due to the variation in the size of masonry blocks and bricks, the sizes of constructed test prisms vary widely. Moreover, as shown in Fig. 4.2b, it seems that not only the *h/t* ratio but also the length of the prism can influence the compressive strength of the prism.

Various studies have been performed to find the effect of *h/t* ratio on the compressive strength of prisms (Maurenbrecher 1980; Fahmy and Ghoneim 1995; Khalaf 1996; Hamid et al. 1985; Hamid and Chukwunenye 1986; Wong and Drysdale 1985). Maurenbrecher (1980) investigated the effect of *h/t* ratio on the compressive strength of concrete clay brick and concrete block prisms having *h/t* ratios of between 1.3 and 5.0. According to the results, for lower *h/t* ratios a higher compressive strength was reported. The same observations were reported by Boult (1979). In another experiment conducted by Brown and Borchelt (1990) on brick prisms having *h/t* ranging between 2.0 and 5.0, it was found that that the compressive strength of hollow concrete prisms (i.e. ungrouted) are less affected by *h/t* than are solid prisms (i.e. grouted or made using solid units). Wong and Drysdale (1985) also performed an experimental study on the strength of prisms having different aspect ratios and concluded that the stress-strain characteristics of prisms made with similar units differ dramatically depending on the direction of compression and whether the prisms are hollow, solid or grouted. According to their report, the compressive strength of grouted prisms decreased from 18.8 MPa in a two-course high prism to 13.0 MPa in a five-course high prism, a drop of about 30%. Also, as the number of courses increased from two to five, the strength decreased accordingly. Moreover, the compressive strength of hollow prisms were found to be nearly the same for three to five course high prisms, and only about 10% higher than the strength of two course high prisms (Wong and Drysdale 1985). This observation for hollow prisms is in contradiction with the result of the finite element study conducted by Fahmy and Ghoneim (1995) in which the strength decreased as the *h/t* ratio increased and remained constant for *h/t* > 5.0. The test results from other researchers also confirmed that the actual unconfined compressive strength value is achieved when the *h/t* ratio is greater than 5.0 (Krefeld 1938; Morel et al. 2007; Walker 2004).

According to another study reported by Hamid et al. (1985) on the effect of *h/t* ratio on the strength of grouted concrete prisms, a distinct mode of failure was observed for prisms with specific number of courses. While for a two-course height prism a shear mode of failure was observed, typical tensile splitting was reported for prisms that consisted of more than two courses. The type of failure is in contrast with the test results from Brown and Borchelt (1990) in which vertical tensile splitting was reported for all brick prisms having *h/t* ratio of between 2.0 and 5.0.

Hamid and Chukwunenye (1986) also conducted three dimensional finite element models on a hollow prism to investigate the effect of h/t ratio. They concluded that a prism needs to consist of more than one mortar joint to represent the actual compressive strength. The same conclusion was reported by Drysdale and Hamid (1979). In another experimental study performed by Khalaf (1996), the strengths of six-course height hollow and grouted prisms were found to be 30 and 10% less than those of two-course hollow and grouted prisms, respectively. Kaaki (2013), conducted an experimental study on 1/3 scaled concrete prisms that consisted of three to five courses with h/t ratios of between 3.28 and 5.52. The five-course height prisms exhibited an average of 13% higher strength than the three-course and four-course height prisms, which was in contrast with previous studies. However, this discrepancy can be attributed to the small scaled size of the tested prisms (Kaaki 2013). Kingsley et al. (1992) also demonstrated that a decrease in prism h/t ratio resulted in an increase in the measured prism compressive strength. Morel et al. (2007) attributed this effect to the confinement effect provided by the loading platens. They showed that the confinement effect decreased as the distance between the loading platens increased.

No study has considered the effect of the length of the prism on the compressive strength. While ASTM C1314 (2003) allows the compressive strength to be obtained using half-block length prisms, no study has investigated the difference between the compressive strength of the full-block length and half-block length prisms. This chapter has examined the effect of the height-to-thickness ratio and the length of the prisms on the compressive strength of concrete masonry prisms.

The primary objectives of the research presented in this chapter are:

- To observe the effect of the frictional force imposed by machine platens at the prism ends
- To study the accuracy of the standards in evaluating the compressive strength of concrete masonry
- To investigate the effect of the length-to-thickness ratio and thickness of the specimens on the compressive strength
- To provide recommendations for strength correction factors of concrete masonry prisms.

4.2 Finite Element Modelling

To perform a finite element study, three dimensional macro models of masonry were simulated using LS-DYNA software. This software is capable of handling dedicated numerical models for non-linear response of concrete/masonry under both static and dynamic loading (LS-DYNA Manual 2007).

In order to mimic the real behavior of masonry prisms, it is imperative that the masonry be modelled as three dimensional solid elements. Therefore, an eight-node brick solid element was used to model the masonry components, having three degrees of freedom in each node. Single point integration was implemented by Gaussian quadrature. This element includes a smeared crack analogy for crushing in compression and cracking in tension. By assigning MAT_CONCRETE_ DAMAGE_REL3, this element is capable of plastic deformation, cracking in three orthogonal directions and crushing. Hourglass control was also provided to avoid the zero energy modes. The masonry constitutive properties are presented and discussed in the next section.

The option BOUNDARY_PRESCRIBED_MOTION was used to induce an increasing displacement to the prisms' ends. This option is helpful to define an imposed nodal motion (velocity, acceleration, or displacement) on a node or a set of nodes. A displacement control of the top nodes, rather than pressure control was used to be able to capture the softening behavior of the material. An explicit time integration solution of LS-DYNA was considered in this study. Sensitivity analyses were conducted to determine the optimum applied displacement rate to minimize the inertial effects. Accordingly, the loading rate of 0.001 mm/s was implemented in the model. In order to mimic the support restraint and confinement pressure due to the end platens, the translational degrees of freedom of nodes were fixed in the lateral directions at the top and bottom faces of the prism model.

4.2.1 Constitutive Masonry Material Model

The size of the member and the level of the confinement affect the properties of the masonry material. A masonry member may be subjected to a simultaneous combination of stresses. The behavior of masonry materials under multi directional stresses is, however, quite challenging to characterize. The inherent similarities between masonry and concrete have tempted many researchers to incorporate calibrated or modified concrete models to simulate the masonry material (Magallanes et al. 2010). However, the response of masonry members can be affected by the mortar layers (Priestley and Elder 1983). In ungrouted masonry, mortar layers are regarded as weak layers vulnerable to cracking, hence, the mode of failure and crack propagation is highly dominated by the mortar layers. Zigzag cracks formed at the mortar layers are a common failure mode of ungrouted masonry members. However, the mortar layers have less influence on grouted masonry members. In terms of applying available concrete material models to masonry members, the lack of coarse aggregate in grouted masonry should be addressed. In concrete materials the strength of coarse aggregate is usually higher than the cement base, which tends to slow down the propagation of the cracks. Lack of coarse aggregates in grout causes the cracks to propagate more freely, hence, the strength degradation and softening branch of the stress-strain curve is steeper in grout compared to concrete. Numerical modelling of masonry also requires a material model that can accurately simulate the volumetric response of masonry material under multi-axial stress, to be able to capture the interaction between inner and outer elements.

The masonry model employed here is K & C concrete damage material model (MAT_CONCRETE_DAMAGE_REL3), which accounts for the complexity of the behavior of concrete. It has been reported that for other similar materials, if an appropriate calibration is provided, the K & C model is able to simulate the behavior and yield excellent results (Magallanes et al. 2010; Ryu et al. 2014). The considered concrete damage material model is a three invariant model in which the failure surface is interpolated between two of three independent shear failure surfaces (LS-DYNA Manual 2007; Magallanes et al. 2010; Malvar et al. 1997; Markovich et al. 2011). These surfaces are the yield surface, maximum surface and residual surface which represent the onset of damage, ultimate and residual strength of the material, respectively.

$$\Delta\sigma_m = a_0 + \frac{P}{a_1 + a_2 P} \quad \text{Yield failure surface} \tag{4.1}$$

$$\Delta\sigma_r = \frac{P}{a_{1f} + a_{2f} P} \quad \text{Maximum failure surface} \tag{4.2}$$

$$\Delta\sigma_m = a_{0y} + \frac{P}{a_{1y} + a_{2y} P} \quad \text{Residual failure surface} \tag{4.3}$$

The constant parameters, a_i, characterize the initial failure surfaces. These surfaces are based on the second invariant of the deviatoric stress:

$$\Delta\sigma = \sqrt{1.5(s_1^2 + s_2^2 + s_3^2)} \tag{4.4}$$

To obtain the shear surfaces for a particular material at least three tri-axial compression tests are required (Schwer and Malvar 2005). For a given confinement pressure, the value of yield, maximum and residual stresses are obtained from compression tests, which are denoted as Pt.1, Pt.2 and Pt.3 in Fig. 4.3a, respectively. Under compression the material response is considered to be linear (the green line in Fig. 4.3a), until it reaches the yield surface (Pt.1). After yielding a strain hardening response governs the behavior before it reaches the maximum strength at Pt.2. The response is then followed by a softening branch until the strength reaches the residual strength (Pt.3). As shown in Fig. 4.3b, these points can be converted to a $\Delta\sigma$ versus P coordinate system and yields a point for each of the shear surfaces. To plot the entire curve, the coordinates of other points of the failure curves are required, which can be obtained through a series of tri-axial tests with different levels of confining pressure.

The tri-axial compression tests reflect the effect of confinement on the behavior and strength of concrete/masonry material. Figure 4.4 indicates the effect of lateral pressure on the concrete stress-strain relationship. The yield, maximum and residual surfaces vary according to the level of lateral pressure.

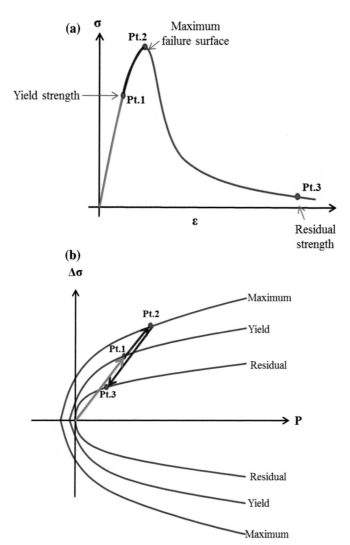

Fig. 4.3 **a** Concrete stress-strain relationship, and **b** three failure surfaces

4.2.2 Damage Function

After the stress reaches the initial yield surface and before it reaches the maximum surface, the current stress surface is determined using a simple linear interpolation between the two surfaces i.e.,

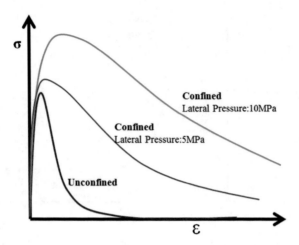

Fig. 4.4 Stress-strain curve for different level of confining pressure (Malvar et al. 1997)

$$\Delta\sigma = \eta\big(\Delta\sigma_m - \Delta\sigma_y\big) + \Delta\sigma_y \qquad (4.5)$$

Similarly, if the current state is located between the maximum and residual surface, the failure surface is defined as:

$$\Delta\sigma = \eta(\Delta\sigma_m - \sigma_r) + \Delta\sigma_y \qquad (4.6)$$

where η is a damage parameter, and represents the relative amount of damage in the current shear surface, varies between 0 and 1, and is a function of accumulated effective plastic strain, λ. The function $\eta(\gamma)$ is a user-defined function of a modified effective plastic strain which is usually determined from experimental data (Malvar et al. 1997).

As shown in Fig. 4.5, the value of $\eta = 0$ at $\lambda = 0$ shows zero plastic strain which is the state before stress reaches the yield surface. The η parameter increases to 1 at $\lambda = \lambda_m$, corresponding to the maximum failure surface, before decreasing back to zero at some greater values of λ. Whenever $\lambda \leq \lambda_m$, the current surface location is between the yield and the maximum failure surface. For $\lambda > \lambda_m$, the state lies between the maximum and residual surfaces.

The damage function is scaled using damage parameters b_1 (compression softening parameter), b_2 (tension softening parameter) and b_3 (tri-axial tension softening parameter) for uni-axial compression, uni-axial tension and tri-axial tension, respectively. The range of the damage scaling parameters and their effect on the behavior is discussed in other references (Magallanes et al. 2010; Malvar et al. 1997; Markovich et al. 2011).

Fig. 4.5 Damage function
(Malvar et al. 1997)

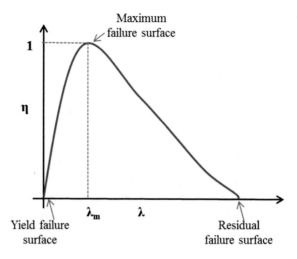

4.2.3 Volumetric Strain

The material model used here for masonry, decouples the volumetric and deviatoric response. The volumetric behavior is governed by a compaction curve or an equation of state (EOS), which relates the pressure, P, to the volumetric strain (Markovich et al. 2011). In addition, EOS also prescribes a set of pressures versus unloading bulk modulus at peak volumetric strains (Noble 2007).

4.2.4 Shear Dilation

In the material model, material expansion due to formation and propagation of cracks can be described by the shear dilation parameter, ω.

The dilation continues as the cracks grow, and stops when the cracks open up enough to clear the aggregates. To consider the effects of shear dilatancy, a proper flow rule must be employed. In the selected material model, a partial associative flow rule is used, prescribed by ω. If $\omega = 0$, no change occurs in volume during plastic flow (non-associative flow without shear dilation) and if $\omega = 1$ shear dilation occurs according to an associative flow rule. A range of $\omega = 0.5$–0.7 is recommended by researchers (Noble 2007). A value of $\omega = 0.5$ is adopted in this study.

4.3 Calibration of Concrete Masonry Prism

Priestley and Elder (1983) conducted an experimental study on concrete masonry prisms. In the experimental study, five-course prisms were constructed from 190 and 140 mm units. In this study, the experimental results on 190 mm prisms reported by Priestley and Elder (1983) are used to calibrate the masonry material model. Unconfined concrete masonry stress-strain curves for the tests performed are presented in Fig. 4.6a (Test 1, Test 2 and Test 3). It can be seen that the behavior is linearly elastic up to the stress level of $0.5f'_m$, followed by a softening behavior before it reaching the peak strength, which approximately corresponds to a strain of 0.0015. The strength degrades rapidly represented by a high slope curve after post peak to a constant value of $0.2f'_m$, for strains greater than ε_{cp}.

Figure 4.6b shows the mesh model used to model the masonry prism. The model consisted of 5120 elements and 6273 nodes, with an average element size of 24.2 mm.

The values of parameters used to model masonry material using the concrete damage model are presented in Table 4.3. Figure 4.7 presents the damage function used in the model calibration process.

Figure 4.6a shows the stress-strain curves under uniaxial stress. Comparing the experimental curves obtained from Test 1, Test 2 and Test 3 with the FEM response, reveals that the stress strain relationship from the finite element analysis is in good agreement with the experimental results of Priestley and Elder (1983). This calibrated model has then been used to perform a parametric study as described below.

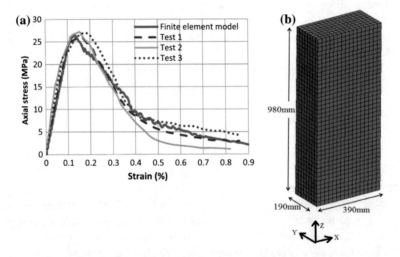

Fig. 4.6 **a** Mesh model, and **b** stress-strain relationship obtained from finite element analysis

Table 4.3 Parameters for masonry material modelling of 26.5 MPa strong masonry	a_0	7.833000	a_{0f}	0
	a_1	0.446300	a_{1f}	0.441700
	a_2	0.003049	a_{2f}	0.004464
	a_{0y}	5.915000	b_1	1.45
	a_{1y}	0.625000	b_2	1.35
	a_{2y}	0.009717	b_3	1.15
			ω	0.5

Fig. 4.7 Damage function of the constitutive masonry model

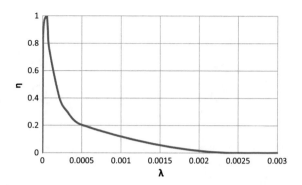

4.4 Parametric Study

The effect of thickness, length-to-thickness and height-to-thickness ratio was explored through a series of parametric studies using finite element modelling. The basic geometry and material properties of the models of the parametric study are similar to the prisms tested by Priestley and Elder (1983) and the calibrated model presented previously (Five-course height prism, with cross section dimensions of 190 mm × 390 mm and a compressive strength of 26.5 MPa)

In the first step of the parametric study, the effect of support confinement was studied. Theoretically, it is expected that by removing the lateral restraints from the top and bottom ends of the prism, the effect of the h/t ratio would be eliminated; i.e. two-course and five-course height prisms would display a similar behavior and exhibit the same load capacity. Figure 4.8a, b show the vertical stress distribution of a frictionless end prism, at the peak strength of five-course and two-course height prisms, respectively. As shown in the figure, in both two-course and five-course prisms the vertical strength of the brick elements are approximately similar (26.4–26.8 MPa), regardless of the location of the element. Low variation of the results numerically confirmed the theoretical concept that if the testing machine does not impose any transverse confinement, the vertical compressive stress in the prism remains uniform within the entire prism. This strength, which is independent of the h/t ratio, characterizes the realistic uniaxial strength of the masonry prism. Comparing the behavior of different elements in the two models also indicates that all elements present approximately the same stress-strain behavior.

Although providing frictionless ends in a compression test is not impossible, generally, in practice, specimens are capped and placed between grooved platens with a degree of friction that imposes lateral confinement at the ends (Malvar et al. 2004). The lateral confinement to the prism at the machine platen interface, results in the ends of the specimen becoming stronger, a non-uniform stress distribution and a barrel deformed shape (Malvar et al. 2004).

Figure 4.9 presents the stress distribution in the vertical direction at the peak strength of five-course and two-course height prisms in which the top and bottom ends are laterally restrained. The prisms are transversely fixed to prevent the end

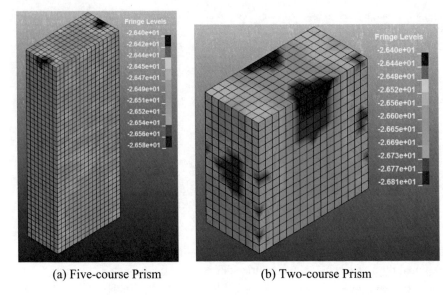

(a) Five-course Prism   (b) Two-course Prism

Fig. 4.8 Stress distribution in vertical direction at the peak strength—frictionless end prisms

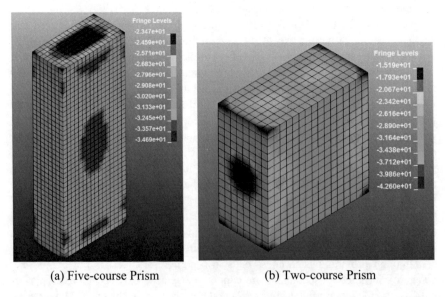

(a) Five-course Prism   (b) Two-course Prism

Fig. 4.9 Stress distribution in vertical direction at the peak strength—with friction at ends

movement to mimic the friction provided by the testing machine. It can be seen that, unlike the stress distribution of the frictionless prisms presented in Fig. 4.8, for both two-course and five-course prisms the vertical stress in the elements varies significantly compared to the frictionless end prisms. While the stress remained relatively constant at about 26.4–26.8 MPa in prisms with frictionless ends, regardless of the number of courses, in prisms with friction the stress of the brick elements ranged between 23.5 and 34.7 MPA and between 15.2 and 42.6 MPa in the five-course and two-course height prism, respectively.

Figure 4.10 presents the vertical stress distribution at the peak strength at the mid-height section of the five-course and two-course prisms with and without friction, presented in Figs. 4.8 and 4.9. In the frictionless prisms, as shown in Fig. 4.10a, b, the stress is approximately constant throughout the cross section (26.4–26.8 MPa) and is independent of the h/t ratio. While there is a wide variation of vertical stresses in the brick element at the mid-height of two-course prism having friction at ends, ranging between 15.9 and 42.6 MPa, the range is comparatively low in the case of the five-course prism, ranging between 24.8 and 26.9 MPa. This is an indicator of the lesser effect of the support friction on the five-course compared with two-course height prisms.

Figure 4.11a indicates the vertical stress versus prism strain of four selected brick elements, M_1 to M_4, located at the mid-height section of the two-course prism with frictionless ends. Figure 4.11b presents the same stress parameter of the same prism but with friction at the ends. It can be seen from Fig. 4.11a that for a prism with frictionless ends, the stress-strain behaviors in different elements are approximately similar, as the elements can expand freely, and are comparatively unconfined.

On the other hand, as indicated in Fig. 4.11b, the same elements exhibited a different response if the ends are restrained, depending on the elements' position in the cross section. As shown in Fig. 4.11b, element M_4 which is a central element, exhibited less strength degradation and developed a higher residual strength compared with other elements. Element M_4, is confined by other elements in both directions. It seems that as the thickness of the prism specimen increases, the confining pressure on internal elements is enhanced.

Figure 4.11 illustrates that the confining pressure on the perimeter elements, which is mainly provided by the block unit elements, is comparatively less than the confining pressure developed in the inner part of the prism, which is due to confinement provided by both grout and block unit elements. Therefore, a material model having unconfined properties similar to a block unit and confined properties of grout can potentially characterize the behavior of masonry. The properties of grout and concrete are similar, as grout is basically a high slump concrete with small aggregates. The main difference is the lack of coarse aggregate in the grout material compared with conventional concrete. Coarse aggregates tend to slow down the crack propagation which results in a higher tensile strength and a lower degradation rate in stress-train curves. Concrete masonry behaves like lightweight concrete in which the coarse aggregate is often weaker than the cement paste, allowing cracks to propagate through the aggregates (Magallanes et al. 2010).

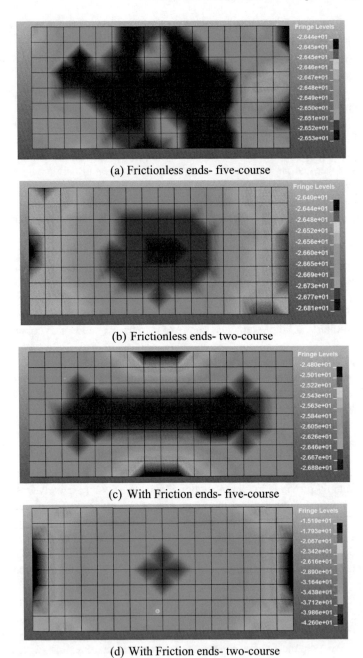

(a) Frictionless ends- five-course

(b) Frictionless ends- two-course

(c) With Friction ends- five-course

(d) With Friction ends- two-course

Fig. 4.10 Vertical stress distribution at the peak strength at mid-height section

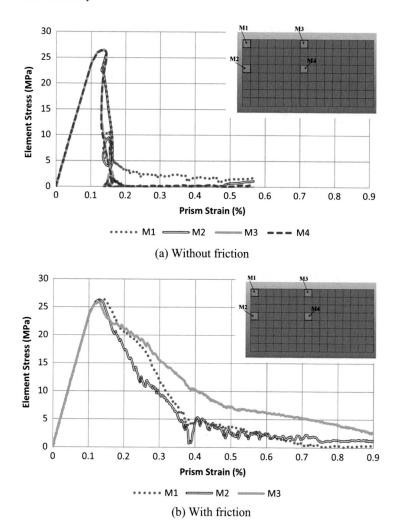

Fig. 4.11 Vertical stress versus prism strain for different elements at the mid-height section of a two-course height prism

4.4.1 Effect of Length

To investigate the effect of block length, 190 mm prisms with full-block length ($l/t = 2.0$), half-block length ($l/t = 1.0$) and double-block length ($l/t = 4.0$) were analyzed and compared.

Figure 4.12a, b presents the stress-strain relationship of full-block and half-block length prisms, respectively. In each figure, prisms comprising different numbers of courses are presented and compared. As can be seen in Fig. 4.12b, in half-block prisms the compressive strength is relatively constant and independent of the number of courses. However, in full-block prisms (Fig. 4.12a), while the compressive strength is nearly similar at the prisms comprising three courses or more, it is comparatively high for the prism with two-course height.

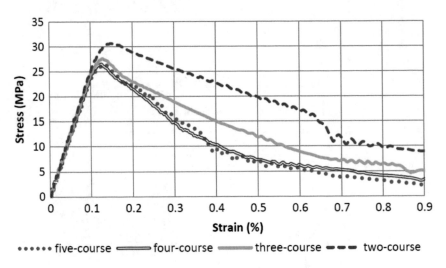

(a) Full-block length prisms

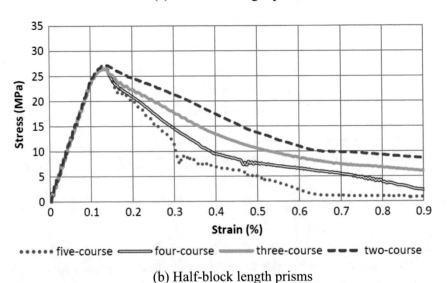

(b) Half-block length prisms

Fig. 4.12 Stress-strain curve of prisms with different lengths and heights

The noticeable difference in strength between a full-block and a half-block prism presented in Fig. 4.12a, b illustrates that not only the height-to-thickness ratio, but also the prism length, affects the maximum compressive strength.

Table 4.4 presents the maximum strength and the relevant correction factors of half, full and double block length 190 mm prisms with different numbers of courses, varying from one to five. The correction factors in the selected masonry codes are also provided to compare. A prism having height-to-thickness ratio of more than five, is considered as a standard prism in which the platen confinement does not affect the strength. According to Table 4.4, the calculated correction factor of full-block and half-block prisms consisting of four courses, is 0.995 and 1.0 respectively. This implies that in order to obtain a realistic stress-strain behavior independent of the top and bottom platen confinement, the height-to-thickness ratio of a full-block and half-block prism should be more than five and four, respectively, beyond which the strength is not influenced by the support restraints. The correction factor of 0.82 in two-course double-block to 0.97 in two-course half-block prisms imply the extensive effect of the prism's length on the strength, which is not reflected in the current versions of masonry codes.

Figure 4.13 compares the longitudinal stress distribution in 190 mm half-block and full-block prisms. The longitudinal stress developed at the end of the specimen represents the lateral confinement at the specimens' ends. As shown in the figure, while in the half-block prism the effect of the confinement is extended up to the first five elements, in the full-block prism the first 11 elements are affected. This indicates why the strength is different in full-block and half-block prisms.

Figure 4.14 presents the stress state at the peak strength in the longitudinal direction of half-block and full-block prisms consisting of different numbers of courses. It can be seen that for h/t of more than two the stress in the longitudinal direction of the half-block prism is nearly constant at the mid-height of the prism. Hence, the effect of confinement due to platen friction is negligible at this section. However, the behavior of the full-block prisms is different. For example, in the two-course full-block prism, the mid-height section is apparently affected by the transverse confinement at the supports; hence, the strength is not actual and is a function of the frictional force. As shown in Fig. 4.14 for h/t of more than five the frictional load does not affect the compressive strength of either full-block or half-block prisms.

Figure 4.15 plots the correction factor vs prism height-to-thickness ratio presented in the selected standards and obtained from the finite element models presented herein. Full-block ($l/t = 2.0$), half-block ($l/t = 1.0$) and double block ($l/t = 4.0$), prisms are considered, to illustrate the effect of the length.

Comparing the correction factors with the ones obtained from the finite element analysis, illustrates that the masonry standards tend to underestimate the strength of half-block prisms, in which the length-to-thickness ratio is about half of the full block. This results in under-prediction of the stiffness of the system, which can cause an unsafe prediction of seismic loads. In half-block prisms with height-to-thickness ratios of greater than 3.0, the strength is nearly unchanged, which is a sign of their independence with respect to the support friction. Generally,

Table 4.4 Correction factor for 190 mm prism

	Height	Height (mm)	Thickness (mm)	Double-block Prism (MPa)	Full-Block Prism (MPa)	Half-block Prism (MPa)	Height-to-thickness ratio	Strength ratio[b] Double-block prism	Full-block prism	Half-block prism	AS4456.4	ASTM C1314 - 03b	CSA S304.1
190 mm-thick	One-course[a]	196	190	53.17	45.16	37.2	1.03	0.53	0.58	0.71	0.70	–	–
	Two-course	392	190	34.41	30.64	27.26	2.63	0.82	0.86	0.97	0.82	0.85	0.86
	Three-course	588	190	31.44	27.63	26.53	3.47	0.89	0.955	0.997	0.885	0.91	0.92
	Four-course	784	190	29.49	26.54	26.42	4.31	0.95	0.995	1.00	0.95	0.96	0.96
	Five-course	980	190	28.07	26.45	26.45	5.15	1.00	1.00	1.00	1.00	1.00	1.00

[a]One-block height prism is presented only to generate the trend, it cannot be used in practice

[b]Strength ratio is the ratio of the strength of the prism to that of the five-block prism

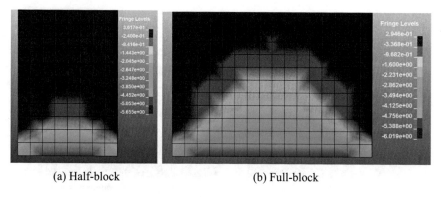

(a) Half-block   (b) Full-block

Fig. 4.13 Stress in longitudinal direction- five-course block

while for prisms with length-to-thickness ratios of more than 4.0, masonry codes yield a reasonable prediction of the strength, for smaller length-to-thickness ratios, as is noticeable from Fig. 4.15, there is a clear error in estimating the strength.

Figure 4.16 compares the normalized strength of 90 and 190 mm prisms for a range of height-to-thickness ratios. The slightly higher slope of the 90 mm prism curves compared with the corresponding curves of 190 mm prisms, illustrates that the strength of a prism with a smaller thickness is more sensitive to the variation of h/t ratio, especially for smaller h/t ratios.

4.5   Discussion and Recommendations for Correction Factor

Figure 4.17 compares the strength correction factors normalized for $h/t = 5.0$. It demonstrates that the correction factors provided in ASTMC1314 are in good agreement for prisms having length-to-thickness ratios higher than 4.0. This is also true for the Canadian standard (CSA S304.1) and Australian Standard (AS4456.4), which are not plotted in Fig. 4.17 to avoid congestion of curves. For smaller length-to-thickness ratios, the masonry codes should consider new correction factors which account for the effect of the length of the prism.

The range of the correction factors obtained in this study for $1.0 < l/t < 4.0$ is presented in Fig. 4.18.

Based on the results of the FEM, two series of correction factors are suggested. In the first series (Table 4.5), prism thickness, l/t ratio and h/t ratio are contributing factors. In the second series (Table 4.6), l/t and h/t are the variables.

Table 4.5 presents recommendations for correction factors for 90 and 190 mm prisms, based on the results of the presented finite element models.

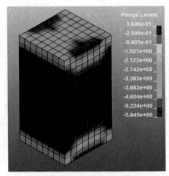

(a) half-block/two-course prism

(b) full block/two-course prism

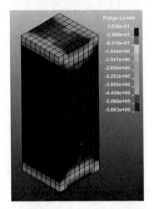

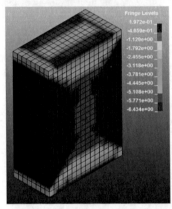

(c) half-block/three-course prism

(d) full block/three-course prism

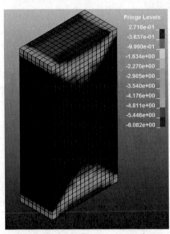

(e) half-block/five-course prism

(f) full block/five-course prism

Fig. 4.14 Stress in longitudinal direction

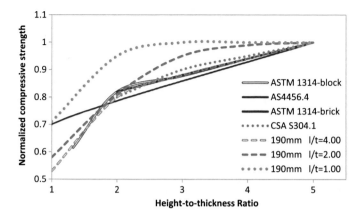

Fig. 4.15 Correction factor for 190 mm prisms

Fig. 4.16 Correction factors
for prism with different
thicknesses

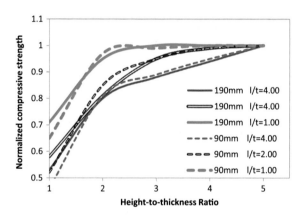

Fig. 4.17 Correction factors
in ASTMC1314 standard
versus FEM

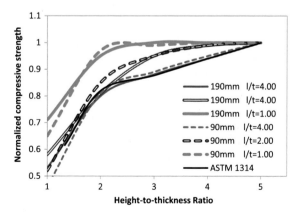

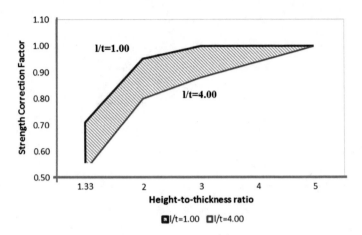

Fig. 4.18 The range of the obtained correction factors

Table 4.5 Recommended correction factors

Prism	l/t	h/t				
		1.00	2.00	3.00	4.00	5.00
Prism thickness = 190 mm	4.00<	0.53	0.80	0.88	0.94	1.00
	2.00	0.58	0.81	0.95	0.99	1.00
	1.00	0.71	0.95	1.00	1.00	1.00
Prism thickness = 90 mm	4.00<	0.46	0.81	0.89	0.95	1.00
	2.00	0.52	0.85	0.95	0.99	1.00
	1.00	0.65	0.97	0.99	1.00	1.00

Table 4.6 Recommended correction factors (simplified)

l/t	h/t				
	1.00	2.00	3.00	4.00	5.00
l/t > 4.00	0.50	0.80	0.88	0.94	1.00
l/t = 2.00	0.55	0.83	0.95	0.99	1.00
l/t = 1.00	0.68	0.96	1.00	1.00	1.00

For prisms having a thickness within the range of 90–190 mm, interpolation can be performed to obtain the correction factors. For example, the correction factor for a full-block length 140 mm thick prism specimen (block size: 140 mm × 190 mm × 39 0 mm), with two-course height, can be determined using interpolation,

$$\frac{h}{t} = \frac{390}{140} = 2.79 \quad and \quad \frac{l}{t} = \frac{390}{140} = 2.79$$

Block	l/t	h/t		
		2.00	2.79	3.00
Prism thickness = 190 mm	4	0.80	0.86	0.88
	2.79		0.896	
	2.00	0.81	0.92	0.95
Prism thickness = 90 mm	4	0.81	0.87	0.89
	2.79		0.906	
	2.00	0.85	0.93	0.95

Hence, for a 140 mm-thick full-block, the strength correction factor can be calculated as:

$$0.896 + \frac{(190 - 140)}{(190 - 90)} * (0.906 - 0.896) \cong 0.9$$

According to Fig. 4.16 there is a slight difference between the correction factors obtained for 90 and 190 mm prisms. Consequently, to simplify the correction factors, the thickness parameter can be ignored by adopting a slight approximation, hence, the correction factors recommended in Table 4.6 can be used instead of the ones presented in Table 4.5.

The correction factors suggested here are based on the finite element models developed in this study for grouted concrete masonry prisms. The same strategy would seem to be necessary to account for the thickness and the length-to-thickness ratio parameters in determining correction factors for other types of masonry prism. An extensive experimental study is also required to verify the effect of the thickness and length-to-thickness ratio in ungrouted/grouted block/brick masonry prisms with varying height-to-thickness ratios.

The rate of the reduction of strength by increasing the h/t ratio is different for prisms with different lengths. The rate is also different in grouted and ungrouted masonry (Khalaf 1996). As the h/t ratio increases, the strength of a grouted prism degrades with a higher rate compared to an identical ungrouted prism having the same geometry.

Figure 4.19a schematically depicts the extent of the stress developed in a hollow prism and Fig. 4.19b, c shows this in a fully grouted full-block and half-block prism having similar h/t ratios and internal material friction angle of φ. As shown, h_c is the height of the pyramid, representing a height of the prism influenced by the support confinement. Comparing Fig. 4.19a, b, reveals that in a hollow prism the confinement effect is limited to the face-shell thickness of the prism. Hence, in hollow prisms, the h_c value is a function of the face-shell thickness, rather than the overall thickness. Consequently, it seems more reasonable to use face-shell thickness to determine the h/t ratio in hollow prisms, as is considered in the Australian standard, AS4456.4 (2003). Comparing Fig. 4.19b, c indicates how the length of the prism affects the h_c of the confined end pyramid. It seems that up to a certain level, as the length of the prism increases, the area influenced by confinement

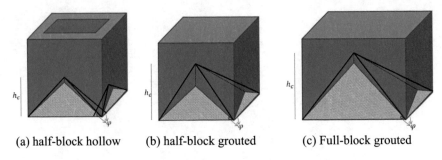

(a) half-block hollow   (b) half-block grouted   (c) Full-block grouted

Fig. 4.19 The height of the wall influenced by the support confinement

extends, hence, h_c increases. This implies that the h/t ratio correction factors in masonry standards need to be revised to account for the effect of length, which can be implemented in the form of h/t or l/t (as recommended in Tables 4.5 and 4.6). It worth mentioning that in the MSJC (2013), f'_m determined using $h/t = 2$, has been used to normalize the nominal strength formulas for masonry members. If a new definition for determining f'_m is introduced, these nominal strength formulas would have to change accordingly.

4.6 Conclusion

This chapter used a numerical finite element analysis to examine the influence of a range of factors on the compressive strength determined through the standard masonry prism tests. The following conclusions were drawn:

– The effect of the friction caused by the end platens in the compression testing machine should not be ignored. It was shown that in the models with frictionless ends the measured strength is the same for specimens regardless of their h/t ratios. On the other hand the effect of friction on increasing the prism strength can be considerable and is a function of h/t ratio.
– The result from the FEM of different sizes of masonry prisms revealed that the compressive strength is not only a function of the thickness but also highly dependent on the length of the prism. In the current masonry codes the effect of the length of the prisms is not considered. As the length-to-thickness ratio increases, the codes' tendency to over-predict the strength becomes more significant. The length-to-thickness ratio is hence an influential parameter in determining the strength of prisms and should be incorporated in the standards' strength correction factors in the future.
– For prisms having equal transverse dimensions, the stress-strain behavior is a function of h/t ratio if this ratio is less than three. In the example presented, in a half-block prism with the cross sectional dimension of 190 mm × 190 mm, the load capacity of three-course block and five-course block was approximately the

same. This indicates that for $h/t > 3.0$ the frictional force due to the machine platen does not affect the strength. In the other example presented, in a full-block prism with the cross sectional dimension of 390 mm × 190 mm, the strength decreased for h/t ratios of up to 5.0.

- The correction factors for concrete and clay masonry prisms are not compatible in the ASTM C1314 (2003) standard. For brick prisms the correction factor is normalized by the strength of a prism having an aspect ratio of 5.0, which seems to represent a realistic strength as the strength seems not to be affected by the machine platen effect. However, according to ASTM C1314 (2003), for concrete masonry prisms the strength is normalized by the prism having a h/t ratio of two. This small aspect ratio seems to be influenced by the support confinement and results in a strength over-prediction of about 20–25%, which is quite considerable and will lead to an unconservative design. To provide a compatible measuring system to obtain the compressive strength independent of the confinement imposed by the machine platen, it is recommended to normalize the correction factors for concrete masonry based on a height-to-thickness ratio of 5.0, the same way as considered in other masonry codes and considered in ASTMC 1314 (2003) for brick masonry. To set the $h/t = 5.0$ as referenced in ASTM C1314, the standard expressions need to be checked and the empirical formulations in the standard need to be revised. For instance the concrete masonry modulus of elasticity $E = 900f'_{m\ new}$ should be altered to $737f'_{m\ new}$ in which $f'_{m\ new}$ is the compressive strength of a concrete prism having an aspect ratio of 5.0.

References

AS 4456.4 (2003) Methods of determining compressive strength of masonry units. Standards Australia, Sydney, NSW, Australia

ASTM C 42-90 (1992) Standard test method for obtaining and testing drilled cores and sawed beams of concrete, ASTM, U.S.A.

ASTM C1314 (2003) Standard test method for compressive strength of masonry prisms. American Society for Testing and Materials, Pennsylvania, United States

Boult B (1979) Concrete masonry prism testing. ACI J Proc ACI 76(4):513–536

Brown RH, Borchelt JG (1990) Compression tests of hollow brick units and prisms. Masonry: components to assemblages, ASTM STP 1063, American society for testing and materials, Philadelphia

CSA S304 (2004) Design of masonry structures. Canadian Standards Association, Mississauga, Canada

Drysdale RG, Hamid AA (1979) Behavior of concrete block masonry under axial compression. ACI J Proc ACI 76(20):1047–1061

Fahmy E, Ghoneim T (1995) Behavior of concrete block masonry prisms under axial compression. Can J Civ Eng 22(5):898–915

Hamid AA, Chukwunenye AO (1986) Compression behavior of concrete masonry prisms. J Struct Eng 112(3):605–613

Hamid AA, Abboud B, Harris H (1985) Direct modeling of concrete block masonry under axial compression. Masonry: research, application and problems. STP-871, ASTM, Philadelphia, 151–166

Kaaki T (2013) Behavior and strength of masonry prisms loaded in compression. Ms thesis, Dalhousie University Dalhousie University, Halifax, Nova Scotia

Khalaf F (1996) Factors influencing compressive strength on concrete masonry prisms. Mag Concr Res 48(175):95–102

Kingsley GR, Noland JL, Schuller MP (1992) The effect of slenderness and end restraint on the behavior of masonry prisms—a literature review. Masonry Soc J 10(2):31–47

Krefeld W (1938) Effect of shape of specimen on the apparent compressive strength of brick masonry. In: Proceedings of the American society of materials, Philadelphia, USA, 363–369

LS-DYNA manual (2007) Version 971. Livermore Software Technology Corporation, Livermore, California, USA, 210–205

Magallanes JM, Wu Y, Malvar LJ, Crawford JE (2010) Recent improvements to release III of the K&C concrete model. In: Proceedings of 11th international LS-DYNA users conference, 6–8 June. Dearborn, MI

Malvar LJ, Crawford JE, Wesevich JW, Simons D (1997) A plasticity concrete material model for DYNA3D. Int J Impact Eng 19(9–10):847–873

Malvar L, Morrill K, Crawford J (2004) Numerical modeling of concrete confined by fiber-reinforced composites. J Compos Constr 8(4):315–322

Markovich N, Kochavi E, Ben-Dor G (2011) An improved calibration of the concrete damage model. Finite Elem Anal Des 47(11):1280–1290

Maurenbrecher A (1980) Effect of test procedures on compressive strength of masonry prisms. Precision second Canadian Masonry symposium, 9–11 June 1980, Ottawa, 119–132

Morel JC, Pkla A, Walker P (2007) Compressive strength testing of compressed earth blocks. Constr Build Mater 21(2):303–309

Noble C (2007) DYNA3D finite element analysis of steam explosion loads on a pedestal wall design. Lawrence Livermore National Laboratory (LLNL), Livermore, California, USA

Priestley M, Elder D (1983) Stress-strain curves for unconfined and confined concrete masonry. ACI J Proc ACI 80(3):192–201

Ryu D, Wijeyewickrema A, ElGawady M, Madurapperuma M (2014) Effects of tendon spacing on in-plane behavior of post-tensioned masonry walls. J Struct Eng 140(4):CID:04013096

Schwer LE, Malvar LJ (2005) Simplified concrete modeling with *MAT_CONCRETE_DAMAGE_REL3., JRI LS-Dyna User Week, Nagoya, Japan

Walker P (2004) Strength and erosion characteristics of earth blocks and earth block masonry. J Mater Civ Eng 16(5):497–506

Wong HE, Drysdale RG (1985) Compression characteristics of concrete block masonry prisms. ASTM Spec Tech Publ 871:167–177

Chapter 5
Flexural Strength Prediction of Unbonded Post-tensioned Masonry Walls

A design equation is developed in this chapter to predict the in-plane flexural strength of unbonded PT-MWs. Using well-validated finite element models, a parametric study is performed to investigate the effect of different parameters on the wall rotation and compression zone length, including axial stress ratio, length and height of the wall, initial to yield stress ratio of PT bars and spacing between PT bars. Multivariate regression analysis is performed to develop an equation to estimate the rotation of the unbonded post-tensioned walls at peak strength. Using the drift capacity of the walls and the proposed equation, a design expression and the relevant step-by-step design method is developed to estimate the flexural strength of unbonded PT-MWs, considering the elongation of PT bars. The proposed design expression is also compared with the predicted values obtained considering no elongation of PT bars which is allowed by the MSJC (2013) standard and validated against experimental results as well as finite element model results.

5.1 Introduction

Masonry is one of the most widely used construction materials in the world. Incorporating post-tensioning into masonry offers a simple and potentially cost-effective structural system. The post-tensioning techniques can be applied to different types of masonry members as either bonded or unbonded reinforcement (Wight et al. 2007; Bean Popehn et al. 2007). Unbonded masonry walls can be ungrouted, partially grouted, or fully-grouted. In the case where grout is used in the cells containing the post-tensioning bars, the post-tensioning (PT) bar is not embedded in the grout and is designed to provide a restoring force to return the wall

A modified version of this chapter has been published in the Journal of Engineering Structures: Hassanli R., ElGawady M. A. and Mills J. E., In-plane flexural strength of unbonded post-tensioned concrete masonry walls, Journal of Engineering Structures, 136, 245–260, 2017.

© Springer International Publishing AG, part of Springer Nature 2019
R. Hassanli, *Behavior of Unbounded Post-tensioned Masonry Walls*,
Springer Theses, https://doi.org/10.1007/978-3-319-93788-5_5

to its original vertical alignment; therefore, reducing residual drifts after a seismic event. In members having bonded PT bars, the force developed in the PT bars can be determined considering strain compatibility and equilibrium. However, for members having unbonded PT bars, the theoretical evaluation of the stress developed in the PT bars is challenging as it depends on the elongation of the unbonded PT bar which in turns depending on the structural element rotations (ElGawady and Sha'lan 2011; ElGawady et al. 2010). Bean Popehn (2007), developed an empirical expression based on finite element analysis of PT-MWs under out-of-plane loads. Wight (2006) used 3D finite element models and developed empirical expressions to determine the strain in PT bars at wall's peak strength. However, the developed finite element models were not able to predict the post-peak performance of the investigated walls. Moreover, Ryu et al. (2014) showed that the expression developed by Wight (2006) were not able to accurately predict the strength of PT-MWs. Similarly, Henry (2012) developed an empirical expression to determine the elongation in PT bar in post-tensioned concrete walls under in-plane loads.

In this study well-calibrated finite element models are used to predict the rotation of the unbonded PT-MWs at peak strength. A parametric study of theoretical walls analyzed using the finite element model was carried out in two stages. In the first stage (set I), the effects of total applied axial stress, initial stress ratio of the PT bars, height of the walls, length of the walls and spacing between PT bars were evaluated on 25 walls. During stage-two (set II), analyses of 45 additional walls were carried out to determine a relationship between the wall rotation and the compression zone length at peak strength, and other parameters including the wall configuration and the level of applied axial stress. An expression was developed based on multivariate regression analysis to predict the rotation and compression zone length at peak strength, which was integrated into a proposed equation to predict the strength of unbonded PT-MWs. The predicted flexural strength using the proposed design equation and the approach allowed by Masonry Standards Joint Committee (MSJC 2013) (No PT bar elongation) were then compared with the values obtained from experimental results available on literature.

5.2 Stress in Unbonded PT Bars

In an unbonded cantilever wall with a flexural mode of failure and a rocking mechanism (Fig. 5.1a), the elongation in PT bar i, is:

$$\Delta l_i = \theta(d_i - c) \tag{5.1}$$

And the stress increment due to elongation Δ_i is:

$$\Delta f_{psi} = \theta \frac{E_{ps}}{L_{ps}}(d_i - c) \tag{5.2}$$

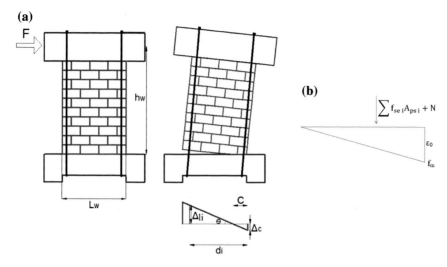

Fig. 5.1 PT-MW **a** before and after deformation, **b** stress distribution at decompression of the heel

Hence, the total stress developed in the ith PT bar can be determined as:

$$f_{psi} = f_{sei} + \theta \frac{E_{ps}}{L_{ps}}(d_i - c) \tag{5.3}$$

where f_{sei} is the effective stress in the ith PT bar after stress losses, L_{ps} is the unbonded length of the PT bar, E_{ps} is the Young's modulus of the pre-stressing steel, c is the compression zone length, and d_i is the distance from the extreme compression fiber to the ith PT bar.

By neglecting the mortar tensile strength, rocking of the wall starts after the wall experiences stresses higher than the decompression stress in the heel. The PT bars display an increase in their initial post-tensioning stresses when the wall-footing interface joint opens. Before this opening, the PT force remains constant, hence:

$$f_{psi} = f_{sei} + (\theta_m - \theta_0) \frac{E_{ps}}{L_{ps}}(d_i - c) \tag{5.4}$$

where θ_m is the wall rotation at peak strength and θ_0 is the rotation corresponding to the decompression point.

Considering a linear stress-strain relationship in the masonry, assuming plane sections remain plane and ignoring the elongation of the PT bars before the decompression point, according to the Fig. 5.1b, the absolute maximum masonry compressive strain corresponding to the decompression point is:

$$\varepsilon_0 = \frac{2(\sum f_{se\,i}A_{ps\,i} + N)}{L_w t_w E_m} \tag{5.5}$$

where A_{ps} is the area of the PT bar(s), E_m is the elastic modulus of the masonry, t_w is the thickness of the wall, and N is the gravity load.

The axial stress, f_m, is defined as:

$$f_m = \frac{\sum f_{se\,i}A_{ps\,i} + N}{L_w t_w} \tag{5.6}$$

Considering a maximum value of 0.15 for f_m/f_m', as recommended by Hassanli et al. (2014a), limits the stress in the masonry corresponding to the decompression point to $0.3\,f_m'$. Hence, considering a linear stress-strain relationship in the masonry at the decompression point is a reasonable assumption.

According to MSJC (2013), the elastic modulus of concrete masonry and clay masonry can be considered as $900f_m'$, and $700f_m'$, respectively. Hence, Eq. 5.5 can be rewritten as:

$$\varepsilon_0 = \left(\frac{1}{450}\right)\frac{f_m}{f_m'} \quad \text{Concrete masonry} \tag{5.7}$$

$$\varepsilon_0 = \left(\frac{1}{300}\right)\frac{f_m}{f_m'} \quad \text{Clay masonry} \tag{5.8}$$

where f_m' is the compressive strength of masonry. The lateral displacement at the top of the wall corresponding to the decompression state is

$$\Delta_0 = \frac{\varphi_0 h_w^2}{3} \tag{5.9}$$

where, φ_0 is the maximum value of the curvature at the decompression point $= \varepsilon_0/L_w$, also $\theta_0 = \Delta_0/h_w$, hence:

$$\theta_0 = \left(\frac{1}{1350}\right)\frac{f_m}{f_m'}\frac{h_w}{L_w} \quad \text{Concrete masonry} \tag{5.10}$$

$$\theta_0 = \left(\frac{1}{900}\right)\frac{f_m}{f_m'}\frac{h_w}{L_w} \quad \text{Clay masonry} \tag{5.11}$$

Using equilibrium, the compression zone length, c, can be expressed as:

$$c = \frac{\sum f_{ps\,i}A_{ps\,i} + N}{\alpha\beta f_m' t_w} \tag{5.12}$$

where α and β are the stress block parameters which are provided by different building codes. (e.g. in MSJC 2013: α = β=0.8).

In order to determine the stress in the tendon using Eq. 5.4, the values of rotation at peak strength, θ_m, and compression zone length at peak strength, c, need to be determined. In this study finite element models are used to determine the relationship between c and θ_m and other parameters of the wall.

5.3 Finite Element Model

Masonry is an anisotropic composite material and therefore modeling depends on the size and material characteristics of the wall elements—the masonry unit, mortar and grout. Masonry structural elements can be modelled using discrete-crack (discontinuum-based) and smeared-crack (continuum-based) approaches (Lourenço 1996). The discrete-crack approach considers the units and mortar joints as separate materials and elements linked by interface relationships. Hence, micro models are very time consuming and computationally demanding. In the smeared-crack approach, blocks, mortar joints, and interfaces are globally represented using a homogeneous material having properties determined from laboratory tests of prisms. All the walls considered in this study were fully grouted, so the effect of the mortar layers on the global behavior is small compared with ungrouped and partially grouted masonry walls (Hassanli et al. 2014a). As the global behavior of the fully grouted walls was of prime importance for this study, a smeared-crack approach was applied and concrete masonry was modelled as a homogenous isotropic material using a nonlinear material model.

Recently, Ryu et al. (2014) developed finite element models for unbonded PT-MWs. However, the model was calibrated on the seismic responses of cavity walls built using clay units. In the current work the model was calibrated and validated against single-leaf PT-MWs built using concrete masonry units (CMU). Three dimensional macro models of masonry walls were simulated using LS-DYNA software, which is a general purpose finite element code. An eight-node brick solid element was used to model the masonry components. This element includes a smeared crack for crushing in compression and cracking in tension. The material model assigned to the masonry elements[1] is capable of plastic deformation, cracking in three orthogonal directions and crushing. The masonry constitutive properties are presented and discussed in the subsequent section. The concrete material for the footing and bond beam was modelled as a linear-elastic material. To simulate the PT bars, two-node beam elements with 2×2 Gauss quadrature and truss formulation were used.

---

[1]MAT_CONCRETE_DAMAGE_REL3.

An elasto-plastic material model with linear kinematic hardening was assigned to PT bar beam elements[2]. The material properties of the PT bar considered in this study were: Young's modulus of 190 GPa, tangent modulus of 2.5 GPa, Poisson's ratio of 0.3 and tensile yield strength of 970 MPa.

Contact elements were used to model the interface between the wall and footing as well as between the wall and bond beam[3]. The assigned contact is capable of simulating interaction between contact interfaces of the two discrete components, which is a significant factor in capturing rocking and sliding behavior. For the stiffness of the surface to surface contact interface, a penalty-based approach was used in which the size of the contact segment and its material properties are used to determine the contact spring stiffness. This method was used as the material stiffness parameters between the contacting surfaces were in the same range. Node-to-surface contact elements[4] were incorporated to model the interface between the PT bars and elements of the footing and bond beam. This contact element prevents the PT bars from penetrating into the solid elements. A soft constraint-based approach was used to calculate the stiffness of the node to surface interface. This approach calculates the stiffness of the linear contact springs based on the nodal masses that come into contact and the global time step size.

An increasing lateral displacement[5] was applied to the nodes at mid-height of the bond beam at the top of the wall. In order to simulate the connection between the footing and the laboratory strong floor, the translational and rotational degrees of freedom of the footing base nodes were fixed.

5.4 Constitutive Model

The basic similarities between masonry and concrete tempt many researchers to use calibrated or modified concrete models to simulate masonry structures (Magallanes et al. 2010). However, mortar joints represent planes of weakness in masonry structures. The effect of these planes of weakness is diminished in fully grouted walls compared to partially grouted and ungrouted walls (Nolph and ElGawady 2012). The lack of large size coarse aggregate in fully grouted masonry is one of its main differences to concrete. The existence of large size aggregate tends to slow down the propagation of the cracks (Magallanes et al. 2010). It has been reported that if an appropriate calibration is provided for masonry, a concrete model such as the one used in this study is able to accurately simulate the behavior of masonry members (Ryu et al. 2014). The K&C (Karagozian & Case) concrete-damage material model that is used in this study was developed by Malvar et al. (1994) and

---

[2]MAT_PLASTIC_KINEMATIC.

[3]AUTOMATIC_SURFACE_TO_SURFACE.

[4]AUTOMATIC_NODE_TO_SURFACE.

[5]BOUNDARY_PRESCRIBED_MOTION.

then improved by Malvar et al. (1996, 1997, 2000). This model can accurately simulate the volumetric response of masonry material under multi-axial stress. The model is described here briefly. More detailed description of the concrete damage model can be found in Malvar et al. (1996, 1997, 2000). The model decouples the deviatoric and volumetric responses.

The deviatoric response includes three invariant models in which the failure surface is interpolated between two of three independent shear failure surfaces (LS-DYNA 2007). As presented in Eqs. 5.13–5.15, these surfaces are functions of hydrostatic pressure, P, and include the yield surface, maximum and residual surface, which represent the onset of damage, ultimate and residual strength of the material, respectively.

$$\Delta\sigma_y = a_{0y} + \frac{P}{a_{1y} + a_{2y}P} \quad \text{(yield failure surface)} \tag{5.13}$$

$$\Delta\sigma_m = a_0 + \frac{P}{a_1 + a_2P} \quad \text{(maximum failure surface)} \tag{5.14}$$

$$\Delta\sigma_r = \frac{P}{a_{1f} + a_{2f}P} \quad \text{(residual failure surface)} \tag{5.15}$$

where P is the mean stress defined as $P = \frac{1}{3}(\sigma_1 + \sigma_2 + \sigma_3)$, $\Delta\sigma_i$ are the failure surfaces for the deviatoric stresses and are functions of the second invariant of the deviatoric stress, J_2 ($\Delta\sigma = \sqrt{3J_2}$), and a_{i-} values are constants defining the failure surfaces and mechanical properties of the material, which can be obtained from experimental tests.

The current stress state, after the stress reaches the yield surface and before it reaches the maximum failure surface (hardening), can be obtained by using linear interpolation between the two surfaces, i.e.,

$$\Delta\sigma = \eta(\Delta\sigma_m - \Delta\sigma_y) + \Delta\sigma_y \tag{5.16}$$

Likewise, the stress state after the stress reaches the maximum failure surface (softening) can be determined by means of linear interpolation between the maximum and residual failure surfaces, i.e.,

$$\Delta\sigma = \eta(\Delta\sigma_m - \Delta\sigma_r) + \Delta\sigma_r \tag{5.17}$$

where η is a damage parameter which represents the extent of damage. η is defined as a function of the effective plastic strain, λ. The full set of the input data of the damage function considered in the masonry material model is provided in Table 5.1.

In the current study, a_i-values were calibrated using test results for masonry prisms as explained in the next section of this chapter.

Table 5.1 Input data of the damage function

λ	η
0	0
8×10^{-6}	0.850
24×10^{-6}	0.970
40×10^{-6}	0.990
56×10^{-6}	1.000
72×10^{-6}	0.990
88×10^{-6}	0.970
550×10^{-6}	0.550
0.002	0.005
0.01	0.005
0.1	0.005
1	0

The volumetric behavior is governed by a volumetric law through a multi-linear compaction model, which relates the pressure P to the volumetric strain. In addition, the volumetric law also prescribes a set of pressures signifying the unloading bulk modulus at peak volumetric strains (Noble 2007). In this study, the pressures and unloading bulk moduli used for masonry are calculated using the following equation (Crawford and Malvar 1997):

$$K = \frac{E_m}{3(1 - 2\upsilon)} \tag{5.18}$$

where υ is Poisson's ratio, and E_m is the elastic modulus, taken as 0.2 and 900 f'm (MSJC 2013) in this study, respectively.

Table 5.2 presents an example of the volumetric law input values for masonry with a compressive strength of 13.3 MPa, considered in the masonry material model. The procedure considered here to determine the material parameters for masonry with a compressive strength of other than 13.3 MPa is provided in Appendix B.

Table 5.2 Volumetric law input data for a 13.3 MPa compressive strength

Volumetric strain	Pressure (MPa)
0	0
−0.0015	10
−0.0043	22
−0.0101	36
−0.0305	68
−0.0513	102
−0.0726	145
−0.0943	221
−0.174	1291
−0.208	1974

5.5 Calibration of the Material Model

In order to determine the parameters which characterize the masonry properties in the considered K&C material model, calibration of masonry prisms was performed. Although the compressive strengths of the masonry walls considered in this study were obtained by testing masonry prisms under compression, the stress-strain curves of the related prisms were not measured. A proper estimate of the masonry constitutive relationship is required for numerical modeling to achieve a reliable global response of a masonry structure. Priestley and Elder (1983) developed a modified Kent-Park (Scott et al. 1982) constitutive model to estimate the stress-strain relationship for confined and unconfined masonry. The proposed stress-strain relationship consists of three portions: a parabolic rising curve (Eq. 5.19), a linear falling branch (Eq. 5.20) and a final horizontal plateau (constant stress) (Eq. 5.21)

$$\varepsilon_m < 0.0015 \rightarrow f_m(\varepsilon_m) = 1.067 f'_m \left[\left(\frac{2\varepsilon_m}{0.002} \right) - \left(\frac{\varepsilon_m}{0.002} \right)^2 \right] \tag{5.19}$$

$$0.0015 \le \varepsilon_m \le \varepsilon_{mp} \rightarrow f_m(\varepsilon_m) = f'_m[1 - Z_m(\varepsilon_m - 0.0015)] \tag{5.20}$$

$$\varepsilon_m > \varepsilon_{mp} \rightarrow f_m(\varepsilon_m) = 0.2 f'_m \tag{5.21}$$

$$\text{where}: Z_m = \frac{0.5}{\left[\frac{3 + 0.29 f'_m}{145 f'_m - 1000} \right] - 0.002} \quad \text{and} \quad \varepsilon_{mp} = \frac{0.8}{Z_m} + \varepsilon_m$$

Figure 5.2a shows a modified Kent-Park stress-strain curve for $f'_m = 13.3$ MPa. In order to calibrate the material model, the same standard-size prism tested by Priestley and Elder (1983) with the thickness/length/height values of 190 mm/ 390 mm/980 mm was modelled in LS-DYNA. Figure 5.2b shows the discretization used to model the masonry prism. The top and bottom nodes were fixed in the transitional x and y direction (Fig. 5.2b), to simulate the confinement induced by the testing machine platen (as discussed in Chap. 4). Similar to the experiment, the top nodes were subjected to increasing axial displacement in the Z-direction (Fig. 5.2b). Figure 5.2a shows the calibrated stress-strain relationship. As shown, the stress-strain curves obtained from the finite element analysis using the K&C material model are in good agreement with the modified Kent-Park model. The values of material parameters obtained as a result of the calibration are not provided here due to space limitations, but can be found in Appendix B.

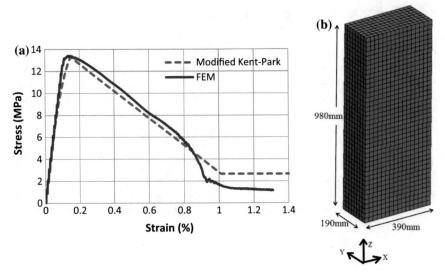

Fig. 5.2 Prism model **a** stress-strain behavior of prism under axial compression load and **b** FEM of prism

5.6 Validation of the Finite Element Model (FEM)

Six fully grouted unbonded PT-MWs tested by Laursen (2002) were selected and analyzed to illustrate the capability and accuracy of the proposed FEM. The walls were tested under cyclic load. The dimensions and configurations of the considered walls are presented in Table 5.3.

These walls were built on a precast reinforced concrete foundation fixed to the laboratory strong floor at the bottom and were free to rotate at top. Accordingly, in the numerical modeling the walls were regarded as cantilevers. During the experimental work the load was applied using an actuator attached to the reinforced concrete (RC) beam fixed to the top of the wall. During the finite element modeling, the RC beam was modelled using a solid element with elastic concrete material and the load was applied to the mid-height of the beam using displacement control. All walls were subjected to in-plane increasing displacement at the top level until failure occurred. Explicit static analysis was used and sensitivity analyses were conducted to determine the optimum applied displacement rate to avoid inertial effects. Densities of 2000 and 2400 kg/m³ were considered in the material model of masonry and concrete, respectively. According to the material testing the PT bar had density, yield strength, tensile strength, elastic modulus and tangent modulus of 8000 kg/m³, 970, 1160 MPa and 190 and 2.5 GPa, respectively, which were implemented in the model. Depending on the wall dimensions, the FEM of the walls consisted of 3767–5375 elements and 5547–7717 nodes and the average element size was 80 mm. Sensitivity analysis was carried out to determine the

Table 5.3 Configuration of the selected experimental walls

Wall*	t (m)	L_w (m)	f'_m (MPa)	No. of PT bars	PT bar initial stress (f_{se}) (MPa)	Initial PT force on the wall (kN)	Axial stress ratio (f_m/f'_m)	V_{EXP} (kN)			V_{FEM} (kN)	V_{EXP}/V_{FEM}	Total PT force at peak strength (kN)		
								Push	Pull	Average			T_{EXP}	T_{FEM}	T_{EXP}/T_{FEM}
W1	0.19	3	13.3	3	468	580	0.077	550	536	543	547	0.99	842	847	0.99
W2	0.14	3	15.1	3	555	690	0.109	472	486	479	460	1.04	776	795	0.98
W3	0.14	3	20.6	2	757	622	0.072	378	368	373	427	0.87	697	689	1.01
W4	0.14	3	17.8	2	757	628	0.084	386	390	388	415	0.94	658	702	0.94
W5	0.14	1.8	20.5	2	534	445	0.086	183	193	188	223	0.84	592	657	0.9
W6	0.14	1.8	18.4	3	614	760	0.164	267	249	258	268	0.96	870	906	0.96

*The typical height of the tested specimens was 2800 mm

element size, according to which further reducing the element size had no effect on the global response of the walls.

The lateral force-displacement obtained from the FEMs and the backbone curves obtained from the experimental cyclic force-displacement response for pull and push directions are plotted in Fig. 5.3. As shown in the figure, the model can predict the wall strength, initial stiffness and rotational capacity with acceptable accuracy. The predicted strength of each test specimen is presented in Table 5.3. As shown in the table, although the FEM tends to over predict the peak lateral load, the V_{EXP}/V_{FEM} ranged from 0.84 to 1.04. The model could approximately predict the stiffness, however, for walls w2, w3 and w4 the stiffness between decompression and

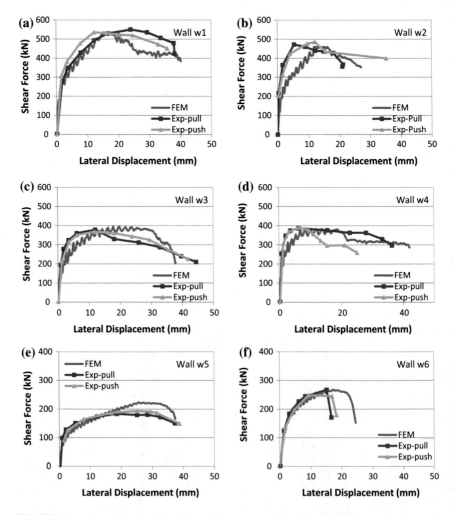

Fig. 5.3 Experimental and calculated force displacement response **a** wall w1, **b** wall w2, **c** wall w3, **d** wall w4, **e** wall w5 and **f** wall w6

yielding is slightly under-predicted (Fig. 5.3). As presented in Table 5.3, comparing the experimental results and the finite element model results indicates that the total PT forces predicted by the FEMs fell within 10% of the experimental results.

The FEM reasonably captured the damage pattern of the PT-MWs. According to the experimental work, a flexural failure mode and gradual strength degradation was observed for all walls (except for wall w6), which was attributed to the spalling of the face shell and crushing of the grout core at the toe zone. The same failure mode was observed in the FEM. Figure 5.4 presents a schematic illustration of damage at the failure stage observed in the experiment and compared with the extent of damage obtained from the finite element analysis. According to the test results, for all walls except wall w6, the plastic deformation was reported to be confined to the lowest two masonry courses, equivalent to 400 mm height. As shown in Fig. 5.4, according to the FEM, the damage is distributed over the first four solid elements at the toe zone, corresponding to a height of 400 mm, which is consistent with the experimental results. Of all the walls, only wall w1 included shear reinforcement at the mid-height. According to the experiment, as shown in Fig. 5.4a, no shear cracks developed in the wall and hence the shear reinforcement was not engaged at all. The FEM result (Fig. 5.4b), shows the same failure and indicates no shear crack in the wall and negligible force developed in the shear reinforcement. Some minor cracks formed at the position of the side PT bars in the test that were not captured in the FEM, which can be attributed to the homogenization of the material used in the FEM. These cracks provided some local effects but they did not influence the global response. Figure 5.4c indicates an inclined crack observed in testing of wall w2, which was attributed to large localized splitting forces of the pre-stress anchorage. The crack did not develop further during testing. As shown in Fig. 5.4d splitting tensile cracks at the location of the PT bar were captured in the FEM. However the model showed a splitting tensile failure along the side PT bars while it occurred along the central PT bar in the test. Of all walls, wall w6 exhibited an unexpected brittle failure mode during the experimental work. The failure mode was characterized by diagonal cracking due to tensile splitting of the masonry compression struts between the post-tensioning anchorage at the top and the toe of the wall. However, wall w6 provided a rocking mechanism and flexural behavior before it reached a drift of 0.5%. Figure 5.5 compares the maximum in-plane stress contour obtained for wall w5 and wall w6 at the peak strength. It is worth noting that Fig. 5.4 shows the damage pattern and plastic contours at the rupture of the walls while Fig. 5.5 shows the stress contours at the peak load. Both wall w5 and wall w6 had the same configuration, but were subjected to different levels of axial stress ratio due to different numbers of PT bars. The stress contours indicate a direct compression strut from the pre-stressing anchorage and toe of wall w6, which did not appear in wall w5. Moreover, according to the force-displacement FEM results, of all considered walls, wall w6 was the only wall which exhibited a brittle failure mode and none of the tensile PT bars yielded, which is consistent with the experimental results.

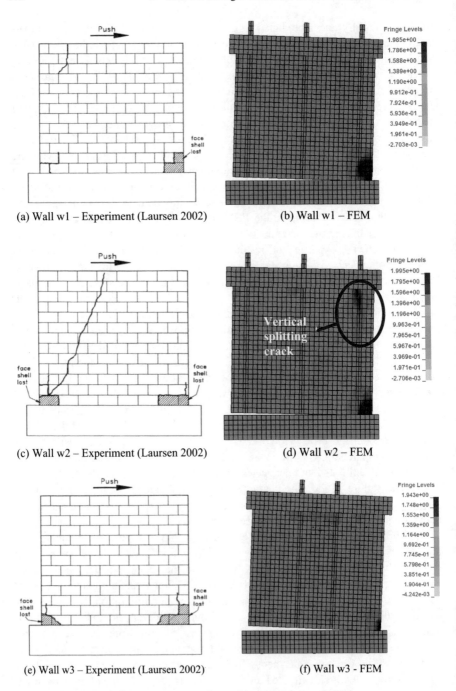

(a) Wall w1 – Experiment (Laursen 2002)   (b) Wall w1 – FEM

(c) Wall w2 – Experiment (Laursen 2002)   (d) Wall w2 – FEM

(e) Wall w3 – Experiment (Laursen 2002)   (f) Wall w3 - FEM

Fig. 5.4 Comparison of damage patterns observed in experiment and FEM

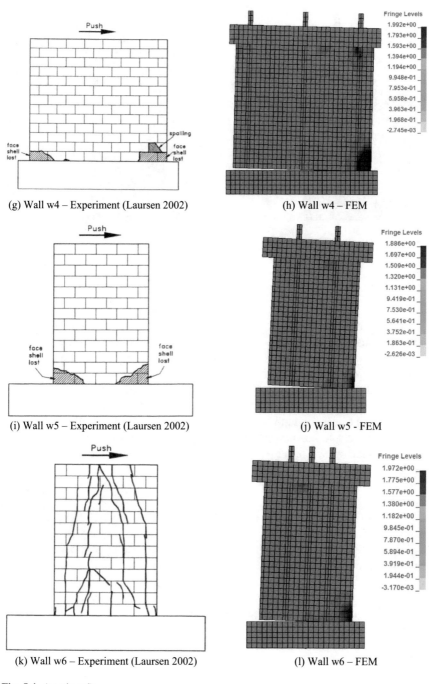

(g) Wall w4 – Experiment (Laursen 2002)   (h) Wall w4 – FEM

(i) Wall w5 – Experiment (Laursen 2002)   (j) Wall w5 - FEM

(k) Wall w6 – Experiment (Laursen 2002)   (l) Wall w6 – FEM

Fig. 5.4   (continued)

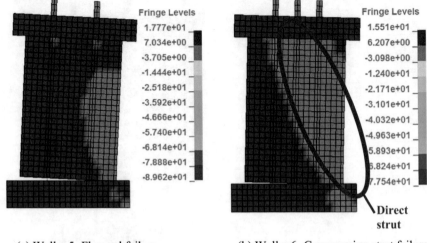

(a) Wall w5- Flexural failure   (b) Wall w6- Compression strut failure

Fig. 5.5 Comparison of maximum in-plane stress contours

Although the FE models correctly predicted the rotation corresponding to the peak strengths of most of the walls, for some of the walls the predicted rotations involved some level of errors. For some walls although similar strengths were recorded in pull and push directions during the experimental work, there was a significant difference in displacements corresponding to the peak strength in pull and push directions. For wall w1, for example, while in the pull direction the displacement at peak strength was 24.07 mm, it was only 12.26 mm in the push direction. The displacement at peak strength obtained from the FEM for Wall w1 was 15.46 mm. Although the FEM model could not capture the rotation at peak strength of walls w2 and w3 accurately, the rotation prediction falls within the range of rotations in the experiment corresponding to strength higher than 90% of the wall strength. Hence, the error in the rotation prediction at peak strength obtained from the FEM is acceptable, as in this study the rotation values at peak strength are only used to predict the wall strength.

These comparisons indicate that the FEMs were able to approximately capture the force-displacement response, maximum strength and failure mode of the PT-MWs and accurately predict the wall drift at the peak strength. The wall drift at peak strength was particularly the interest of FEMs, as it is used in this paper to develop design expression to predict the stress developed in the PT bars and to calculate the lateral strength of the PT-MWs. Therefore, the developed FEM was adopted for a parametric study to investigate the effect of different parameters on the behavior of PT-MWs.

5.7 Parametric Study

The validated finite element model was then used to conduct a parametric study to develop expressions for the rotation and flexural strength of PT-MWs. The parametric study was performed in two stages by considering two different sets of wall models.

The first set was developed to evaluate the effect of different parameters on the flexural strength and drift capacity of walls at peak strength. Based on the analysis results obtained from this set, design recommendations were provided for PT-MWs. Subsequently, matrices of the wall models were defined for the second set of the parametric study. Following this step, expressions for rotation and a design method to evaluate the flexural strength of unbonded PT-MWs were developed.

5.7.1 Parametric Study-Set I

The first stage of the parametric study included 25 walls in five groups (Table 5.4) to assess the influence of different parameters on strength and deformation. As presented in Table 5.4, the walls W1-1 to W1-5, W2-1 to W2-8, W3-1 to W3-4, W4-1 to W4-5 and W5-1 to W5-3 were used to study the effects of f_m/f'_m, f_i/f_{py}, height of the wall, spacing of the PT bars and the length of the wall, respectively, on the strength and deformation of the wall

A wall with the dimensions height/thickness/length of 2800 mm/190 mm/3000 mm was considered as the "control" specimen. The compressive strength of the masonry and the thickness of the wall in both sets of the parametric study were taken as 13.3 MPa and 190 mm, respectively.

5.7.1.1 Effect of Axial Stress Ratio f_m/f'_m

Five different values of axial force of 204.8, 577.5, 831.0, 1060.5 kN and 2112.0 corresponding to 2.7, 7.6, 11.0, 14.0 and 27.9% of f'_m were considered for walls W1-1 to W1-5, respectively. Figure 5.6a presents the base shear versus displacement obtained for these walls from the FEM. As shown in the figure, as the f_m/f'_m ratio increases, the strength increases while the ductility decreases. Walls with higher ratios of f_m/f'_m experienced a sudden degradation trend beyond the peak strength. Figure 5.6b illustrates the effect of the axial stress ratio on the lateral strength of PT-MWs.

Although an increase in f_m/f'_m results in an increase in the strength of the wall, the rate of strength increase varies depending on the level of axial stress ratio. Strength increase due to an increase in the axial stress ratio is more pronounced until f_m/f'_m reaches a value of 0.14. Beyond this value, the axial stress ratio has a relatively small effect on the in-plane strength. It is worth mentioning that based on

Table 5.4 Wall matrix of set I- parametric study[*]

Variable	Wall	h (mm)	t_w (mm)	L_w (mm)	f_m/f'_m	No. of PT bars	f_{py} (MPa)	f_{pu} (MPa)	Initial stress, f_{se} (MPa)	Axial Force (kN)	f_i/f_y
f_m/f'_m	W1-1	2800	190	3000	0.027	3	970	1160	495	204.8	0.51
	W1-2	2800	190	3000	0.076	3	970	1160	466	577.5	0.48
	W1-3	2800	190	3000	0.110	3	970	1160	447	831.0	0.46
	W1-4	2800	190	3000	0.140	3	970	1160	428	1060.5	0.44
	W1-5	2800	190	3000	0.279	3	970	1160	352	2112.0	0.36
f_i/f_y^{**}	W2-1	2800	190	3000	0.076	3	1595	1907	466	577.5	0.29
	W2-2	2800	190	3000	0.076	3	970	1160	466	577.5	0.48
	W2-3	2800	190	3000	0.076	3	709	848	466	577.5	0.66
	W2-4	2800	190	3000	0.075	3	555	663	459	568.5	0.83
f_i/f_y^{***}	W2-5	2800	190	3000	0.076	3	970	1160	466	577.5	0.29
	W2-6	2800	190	3000	0.076	3	970	1160	466	577.5	0.48
	W2-7	2800	190	3000	0.076	3	970	1160	466	577.5	0.66
	W2-8	2800	190	3000	0.075	3	970	1160	459	568.5	0.83
h	W3-1	1800	190	3000	0.074	3	970	1160	454	562.5	0.47
	W3-2	2800	190	3000	0.076	3	970	1160	466	577.5	0.48
	W3-3	3800	190	3000	0.077	3	970	1160	472	585.0	0.49
	W3-4	4800	190	3000	0.078	3	970	1160	478	592.5	0.49
Spacing	W4-1	2800	190	5000	0.068	2	970	1160	414	856.0	0.43
	W4-2	2800	190	5000	0.073	3	970	1160	446	921.0	0.46
	W4-3	2800	190	5000	0.074	4	970	1160	451	932.0	0.47
	W4-4	2800	190	5000	0.052	5	970	1160	476	655.0	0.49
	W4-5	2800	190	5000	0.069	8	970	1160	477	876.0	0.49
Length	W5-1	2800	190	1800	0.077	3	970	1160	468	348.0	0.48
	W5-2	2800	190	3000	0.076	3	970	1160	466	577.5	0.48
	W5-3	2800	190	4200	0.078	3	970	1160	456	824.0	0.47

[*]Typical unbonded length = h + 800 mm, typical PT bar diameter ranges from to 13 to 50 mm
[**]f_y variable, f_i constant
[***]f_i variable, f_y constant

analysis of experimental results of 31 walls specimens, Hassanli et al. (2014b) recommended an upper limit of f_m/f'_m of 0.15. Moreover, a higher value of f_m/f'_m results in a lesser ductility value. Figure 5.7a, b presents the damage pattern at the peak strength of the walls W1-4 and W1-5. As shown in Fig. 5.7a, while for the wall W1-4 having an axial stress ratio of 0.14 a flexural mode of failure is observed, by increasing the axial stress ratio to 0.278 in the wall W1-5, a brittle splitting mode of failure controls the response, as indicated in Fig. 5.7b. This is consistent with the results reported by Laursen (2002) and Hassanli et al. (2014b) that a brittle failure might occur by increasing the axial stress ratio. The sudden degradation beyond the

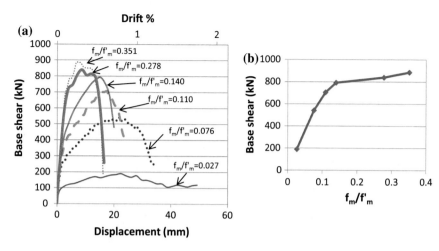

Fig. 5.6 Effect of axial stress ratio on **a** force-displacement response, and **b** lateral strength

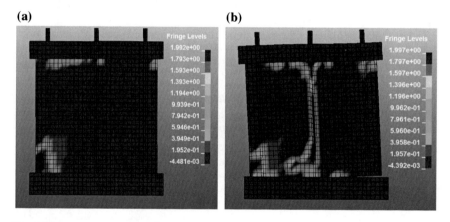

Fig. 5.7 Damage pattern **a** wall W1-4, **b** wall W1-5

peak strength for walls with high axial stress ratios can be attributed to the effect of high axial stress on changing the mode of failure from ductile to brittle failure. This is in line with the explanation presented previously for walls w5 and w6 in Fig. 5.5. It therefore seems that the level of axial stress ratio has a significant effect on the strength, ductility, behavior and failure mode of PT-MWs. Hence, it is recommended to limit the axial stress ratio to a value of 0.15. Moreover, as presented in Fig. 5.6b, for $f_m/f'_m < 0.15$, the strength of the wall is highly dependent on the level of the axial stress ratio. Therefore, to provide both adequate strength and economical design in a PT-MW, a small axial stress ratio, for example less than 0.05 should be avoided.

5.7.1.2   Effect of the Initial to Yield Stress Ratio of PT Bars f_i/f_{py}

To investigate the effect of the initial to yield stress ratio in the PT bars, walls W2-1 to W2-8 were generated. For these walls all parameters including the number of PT bars and the wall configuration were made the same, and the only variable was f_i/f_{py}. Four different values of f_i/f_{py} = 0.29, 0.48, 0.66, 0.83 were considered. To provide the same axial stress ratio for walls W2-1 to W2-4, the total initial PT force was kept the same; however the yield strength of the PT bars was varied, as presented in Table 5.4. To provide the same axial stress ratio for walls W2-5 to W2-8, the total initial PT force was kept the same by considering different cross-sectional areas of PT bars; however the initial stress of the PT bars was varied.

Figure 5.8 shows the base shear versus displacement obtained for this group of walls. As shown in the figure, a higher level of f_i/f_{py} resulted in a smaller strength and a greater ductility compared to walls having smaller values of f_i/f_{py}. This can be explained by the force developed in the PT bars during deformation. While in wall W2-4 and W2-8 a large value of f_i/f_{py} = 0.83 caused the PT bars to yield at a small drift value of about 0.12%, in walls W2-1 and W2-5 with a small value of f_i/f_{py} = 0.29 higher values of PT force were developed prior to yielding, resulting in a higher lateral strength of the wall.

Figure 5.9a, b compares the total force developed in the PT bars of walls W2-1 to W2-4 and walls W2-5 to W2-8, respectively. According to Fig. 5.9a, while the total initial PT forces were the same for all walls, the total forces reached a maximum of 982.6, 910.1, 763.9 and 648.9 kN, corresponding to an increase of 70.3, 57.7, 32.4, and 12.5% of the initial PT force of the walls W2-1 to W2-4, respectively. Moreover, according to Fig. 5.9b, the total PT forces in walls W2-5 to W2-8 reached a maximum of 1096.2, 910.1, 775.2 and 656.5 kN, corresponding to an increase of 90.0, 57.7, 34.3, and 13.8% of the initial PT force, respectively.

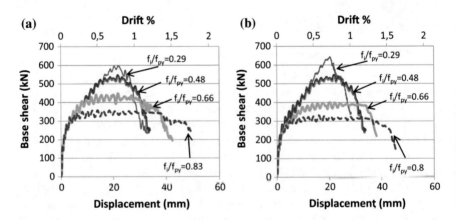

Fig. 5.8 Effect of initial to yield stress ratio in PT bars **a** f_i constant, f_{py} variable and **b** f_i variable, f_{py} constant

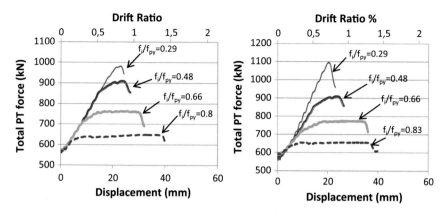

Fig. 5.9 Effect of initial to yield stress ratio on the total PT force **a** f_i constant, f_y variable and **b** f_i variable, f_y constant

Hence, while it is not cost-effective to keep the f_i/f_{py} ratio very small, since a larger bar area is required to provide the same post-tensioning force, a high value of f_i/f_{py} is also not economical as the PT bars yield at a small deformation and the wall cannot develop higher strength. The rotational capacity of the walls at the maximum strength was used in this study in order to determine the flexural strength. For a high level of f_i/f_{py}, a range of θ values can be considered for the rotational capacity. For example, according to Fig. 5.8a, for wall W2-4, any displacement value between 8 and 37 mm corresponds to the maximum strength and can be considered to calculate the rotational capacity (drift ratio). To avoid such a wide range of values in a specific wall and also to prevent premature failure of the wall due to yielding of the PT bars at small displacements, the f_i/f_{py} ratio was limited to 0.6 in generating the wall configuration for wall set II of the parametric study. It is worth noting that it is common practice in the U.S to limit the post-tensioning force, after initial losses, to 45–65% of f_{py}. Hence, this recommendation is in line with current practice.

5.7.1.3 Effect of the Height

To investigate the effect of the height on the behavior of PT-MWs, walls W3-1 to W3-4 having heights ranging from 1800 to 4800 mm were investigated (Table 5.4). Figure 5.10a presents the base shear versus displacement results for these walls. As shown in the figure, while by increasing the height of the wall the displacement capacity increases, the base shear capacity decreases. To normalize the response, the base moment versus drift curve is presented in Fig. 5.10b. As shown in the figure, the peak moment strength and the drift corresponding to the peak strength is approximately the same for all walls, regardless of the height of the wall. Consequently, within the range of the parameters considered in this study, it seems that the drift capacity of the wall at peak strength is not a function of the height of the wall.

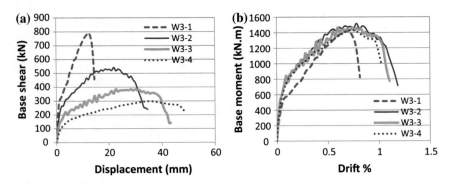

Fig. 5.10 **a** Force-displacement curve, and **b** moment-drift curve for walls of set I, having different heights

Figure 5.11 presents the base crack profile of the walls W3-1 to W3-4 at the peak strength. As shown in the figure, the compression zone length is approximately the same for all considered walls. The compression zone length, c, for walls W3-1, W3-2, W3-3 and W3-4 was found to be equal to 349.7, 353.0, 354.4 and 363.2 mm, corresponding to 11.66, 11.77, 11.81 and 12.10% of L_w, respectively. The variation is within two percent of the average value of c and can therefore be ignored. Hence, the compression zone length, c, can be considered to be independent of the height of the wall.

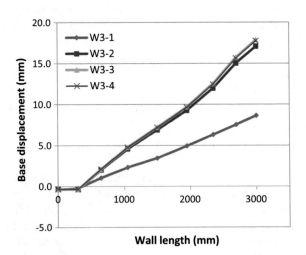

Fig. 5.11 Base crack profile

5.7.1.4 Effect of the PT Bar Spacing

As presented in Table 5.4, Walls W4-1 to W4-5 were generated to investigate the influence of the spacing between PT bars. As shown in Table 5.4, the number of PT bars considered in walls W4-1 to W4-2 were, two, three, four, five and eight, corresponding to spacing of 4400, 2200, 1400, 1100, and 600 mm, respectively. Figure 5.12 compares the damage pattern of the walls at the maximum strength. Comparing the damage pattern reveals that while in the walls with spacing of more than 1400 mm, corresponding to $7t_w$, vertical splitting cracks develop in the compression zone at the location of the PT bar, a flexural failure mode is exhibited by wall W4-2 in which the PT bars are spaced closely. The same mode of failure for widely spaced PT bars has also been reported by other researchers (Ryu et al. 2014). According to Fig. 5.12, to avoid vertical splitting cracks, which develop close to the PT bar on the compression side, the PT bar spacing should be limited to a value between 1100 and 1400 mm. Considering a spacing of six times the wall nominal thickness ($6 \times 200 = 1200$ mm in this study) as the maximum spacing, seems to yield an acceptable limit to prevent vertical splitting cracks from occurring. This limit is similar to that required by MSJC (2013) for out-of-plane loads. As lateral loads are multi-directional, structural walls need to be designed for both in-plane and out-of-plane loading. Hence, for set II of the parametric study, the maximum spacing between PT bars was considered to be equal to $6t_w$ to satisfy the standard limitations on out-of-plane loading.

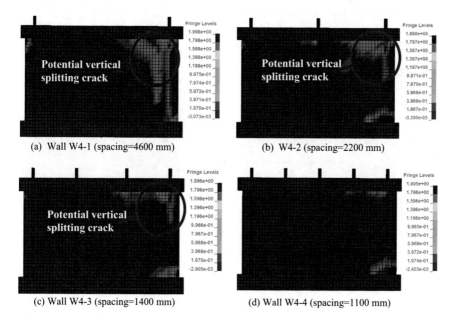

(a) Wall W4-1 (spacing=4600 mm)   (b) W4-2 (spacing=2200 mm)

(c) Wall W4-3 (spacing=1400 mm)   (d) Wall W4-4 (spacing=1100 mm)

Fig. 5.12 The effect of spacing on failure mode and damage pattern

5.7.1.5   Effect of Wall Length

Walls W5-1 to W5-3 with lengths of 1800, 3000 and 4200 mm respectively, were considered to investigate the effect of the wall length on the drift and flexural strength (Table 5.4). Figure 5.13a compares the force vs displacement curves of the walls with different lengths, and Fig. 5.13b presents the base shear normalized by $f'_m A_n$ versus drift. As shown in the figure, by increasing the wall length while the strength increases, the ductility reduces. Consequently, the wall length has an influential effect on the drift capacity, strength and general behavior of the PT-MWs.

5.7.2   Parametric Study-Set II

The configuration of the walls in set II of the parametric study was determined according to the conclusions obtained from set I of the parametric study. Table 5.5 presents the wall matrix of set II of the parametric study. The yield strength, elastic modulus and tangent modulus were set to 970 MPa, 190 and 2.5 GPa, respectively for all PT bars, the thickness of all walls was 190 mm, the maximum spacing between PT bars was six times the wall thickness and the f'_m of all walls was 13.3 MPa. As shown in Table 5.5, three different lengths of 1800, 3000 and 4200 mm and three different heights of 2800, 3800 and 4800 mm were considered in this set for analysis, the axial stress ratio was varied from 0.05 to 0.15%, and the f_i/f_{py} value ranged from 0.3 to 0.6.

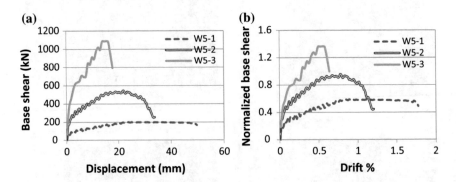

Fig. 5.13 **a** Force-displacement curve and **b** normalized based shear versus drift of walls of set I, having different lengths

Table 5.5 Wall matrix of set II- parametric study

Wall name	h (m)*	L_w (m)	Self-weight (kN)	f_m/f'_m	No. of bars*	L_{ps} (mm)	Initial stress, f_{se} (Mpa)	Total pre-stressing force (kN)	θ_m%	c (m)	θ_0%
W1.8-2.8-1	2.8	1.8	23	0.052	2	2.6	482.3	236	0.86	0.18	6.00E-03
W1.8-2.8-2	2.8	1.8	23	0.077	2	2.6	467.9	348	0.92	0.18	9.00E-03
W1.8-2.8-3	2.8	1.8	23	0.098	2	2.6	455.8	446	0.89	0.26	1.00E-02
W1.8-2.8-4	2.8	1.8	23	0.12	2	2.6	444.7	544	0.89	0.38	1.00E-02
W1.8-2.8-5	2.8	1.8	23	0.14	2	2.6	433.3	636	0.93	0.39	2.00E-02
W1.8-3.8-1	3.8	1.8	31	0.052	2	2.6	486.4	238	1.22	0.16	8.00E-03
W1.8-3.8-2	3.8	1.8	31	0.078	2	2.6	476	354	1.15	0.22	1.00E-02
W1.8-3.8-3	3.8	1.8	31	0.1	2	2.6	463.9	454	1.02	0.32	2.00E-02
W1.8-3.8-4	3.8	1.8	31	0.122	2	2.6	454.6	556	1.02	0.39	2.00E-02
W1.8-3.8-5	3.8	1.8	31	0.144	2	2.6	445.5	654	0.87	0.37	2.00E-02
W1.8-4.8-1	4.8	1.8	39	0.053	2	2.6	489.5	239.5	1.24	0.16	1.00E-02
W1.8-4.8-2	4.8	1.8	39	0.078	2	2.6	478.7	356	1.16	0.26	2.00E-02
W1.8-4.8-3	4.8	1.8	39	0.101	2	2.6	470.1	460	1.03	0.31	2.00E-02
W1.8-4.8-4	4.8	1.8	39	0.124	2	2.6	461.1	564	1.03	0.39	2.00E-02
W1.8-4.8-5	4.8	1.8	39	0.146	2	2.6	452.3	664	0.83	0.36	3.00E-02
W3.0-2.8-1	2.8	3	38	0.052	3	3.8	533.4	391.5	1.07	0.28	4.00E-03
W3.0-2.8-2	2.8	3	38	0.076	3	3.8	517.6	577.5	0.68	0.4	5.00E-03
W3.0-2.8-3	2.8	3	38	0.097	3	3.8	502.8	738	0.58	0.59	7.00E-03
W3.0-2.8-4	2.8	3	38	0.119	3	3.8	490.5	900	0.58	0.59	8.00E-03
W3.0-2.8-5	2.8	3	38	0.139	3	3.8	476.9	1050	0.55	0.68	1.00E-02
W3.0-3.8-1	3.8	3	52	0.052	3	3.8	539.5	396	1.08	0.29	5.00E-03
W3.0-3.8-2	3.8	3	52	0.077	3	3.8	524.9	585.6	0.64	0.36	7.00E-03
W3.0-3.8-3	3.8	3	52	0.099	3	3.8	513	753	0.54	0.5	9.00E-03

(continued)

Table 5.5 (continued)

Wall name	h (m)*	L_w (m)	Self-weight (kN)	f_m/f'_m	No. of bars*	L_{ps} (mm)	Initial stress, f_{se} (Mpa)	Total pre-stressing force (kN)	$\theta_m\%$	c (m)	$\theta_0\%$
W3.0-3.8-4	3.8	3	52	0.121	3	3.8	500.3	918	0.54	0.59	1.00E-02
W3.0-3.8-5	3.8	3	52	0.142	3	3.8	489.1	1077	0.53	0.69	1.00E-02
W3.0-4.8-1	4.8	3	66	0.053	3	3.8	543.6	399	0.9	0.27	6.00E-03
W3.0-4.8-2	4.8	3	66	0.078	3	3.8	530.3	591.6	0.64	0.4	9.00E-03
W3.0-4.8-3	4.8	3	66	0.101	3	3.8	519.5	762.6	0.56	0.49	1.00E-02
W3.0-4.8-4	4.8	3	66	0.123	3	3.8	508.8	933.6	0.56	0.59	1.00E-02
W3.0-4.8-5	4.8	3	66	0.145	3	3.8	498.7	1098	0.51	0.68	2.00E-02
W4.2-2.8-1	2.8	4.2	54	0.054	4	5	582.4	570	0.72	0.47	3.00E-03
W4.2-2.8-2	2.8	4.2	54	0.078	4	5	553.9	824	0.59	0.79	4.00E-03
W4.2-2.8-3	2.8	4.2	54	0.096	4	5	523.2	1024	0.47	0.89	5.00E-03
W4.2-2.8-4	2.8	4.2	54	0.121	4	5	523.2	1280	0.47	0.95	6.00E-03
W4.2-2.8-5	2.8	4.2	54	0.139	4	5	504.1	1480	0.46	1.38	7.00E-03
W4.2-3.8-1	3.8	4.2	73	0.054	4	5	588.6	576	1.43	0.46	4.00E-03
W4.2-3.8-2	3.8	4.2	73	0.079	4	5	564.7	840	1.12	0.59	5.00E-03
W4.2-3.8-3	3.8	4.2	73	0.099	4	5	537.5	1052	0.93	0.78	7.00E-03
W4.2-3.8-4	3.8	4.2	73	0.124	4	5	539.6	1320	0.93	0.88	8.00E-03
W4.2-3.8-5	3.8	4.2	73	0.145	4	5	525.9	1544	0.85	0.97	1.00E-02
W4.2-4.8-1	4.8	4.2	92	0.055	4	5	596.8	584	1.53	0.39	5.00E-03
W4.2-4.8-2	4.8	4.2	92	0.08	4	5	572.8	852	1.17	0.62	7.00E-03
W4.2-4.8-3	4.8	4.2	92	0.101	4	5	545.7	1068	0.89	0.77	9.00E-03
W4.2-4.8-4	4.8	4.2	92	0.127	4	5	552.7	1352	0.89	0.89	1.00E-02
W4.2-4.8-5	4.8	4.2	92	0.149	4	5	539.5	1584	0.84	0.96	1.00E-02

*Typical unbonded length = h + 800 mm
*Typical PT bar diameter ranges from to 25 to 61.16 mm

5.8   Proposed Expression to Predict the Flexural Strength of Unbonded Masonry Walls

The wall's rotation at peak strength, θ_m, and the compression zone length at peak strength, c, obtained from the finite element analysis are presented in Table 5.5. According to the results, the drift ratio, θ, ranged from 0.46 to 1.53%, and the compression zone length, c, ranged from 160 to 1375 mm, corresponding to 9.7 and 32.7% of the wall length, L_w, respectively.

Different factors contribute to the rotation and the compression zone length at peak strength including: length, height and thickness of the wall, axial stress ratio, longitudinal and shear reinforcement ratio, steel, concrete/masonry material properties, level of confinement, loading pattern and moment gradient. Considering the walls failed in flexure and using the values obtained for θ_m and c for the different wall configurations (Table 5.5), a multivariate regression analysis was carried out. As a result, the wall's length and the axial pre-stress ratio were found to be the most influential factors affecting θ_m and c. Considering these two parameters, the following equation was obtained from the multivariate regression analysis to estimate the rotation of unbonded PT-MWs at their peak strength:

$$\theta_m c = \left(0.00055 L_w + 17.375 \frac{f_m}{f'_m}\right) \tag{5.22}$$

The obtained values for R^2 (coefficient of determination) and adjusted-R^2 (adjusted coefficient of determination) were both higher than 90%. Design of experiments (DOE) of data showed a low interaction between the parameters. The application of this expression is limited to wall members having a flexural mode of failure.

Figure 5.14a, b indicate the effect of the wall length and axial stress ratio on the $\theta_m c$ of set II of walls. The upward trend of data presented in Fig. 5.14a reveals that as the wall length increases the $\theta_m c$ value increases. Moreover, according to Fig. 5.14b and Eq. 5.22, the $\theta_m c$ value for members having a higher level of axial stress ratio is larger. According to Fig. 5.1a, considering a constant value for ε_{mu}, the plastic hinge length, L_{pl}, is proportional to $\theta_m c$. ($\theta_m c = \varepsilon_{mu} L_{pl}$).

Substituting Eq. 5.22 into Eq. 5.4, the force developed in the PT bar at peak strength can be expressed as

$$f_{psi} = f_{sei} + \left[\frac{0.00055 L_w + 17.375 \frac{f_m}{f'_m}}{c} - \theta_0\right] \frac{E_{ps}}{L_{ps}} (d_i - c) \tag{5.23}$$

Using Eq. 5.10, the values of θ_0 for the walls of set II of the parametric study were obtained and presented in Table 5.5. As presented in the table, the value of θ_0 is considerably smaller than θ_m, and hence can be ignored in comparison with θ_m.

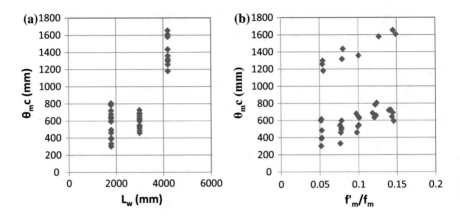

Fig. 5.14 Effect of the **a** length and **b** axial stress ratio on $\theta_m c$

Therefore, the following Equation is proposed to predict the stress developed in an unbonded PT-MW at peak strength:

$$f_{psi} = f_{sei} + \left(0.00055L_w + 17.375\frac{f_m}{f'_m}\right)\frac{E_{ps}}{L_{ps}}\left(\frac{d_i}{c} - 1\right) \leq f_{py} \qquad (5.24)$$

The following steps can be followed to estimate the strength of a PT-MW:

(1) Consider $f_{ps} = f_{se}$,
(2) Determine the compression length, c, using Eq. 5.12,
(3) Determine the stress in the tendons, f_{psi}, using Eq. 5.24,
(4) Perform an iteration of steps 2 and 3 to reach a constant value of c,
(5) Determine the moment capacity using Eq. 5.25 and using the c and f_{psi} values obtained in step 4.

$$M = \sum f_{psi}A_{psi}\left(d_i - \frac{a}{2}\right) + N\left(\frac{L_w}{2} - \frac{a}{2}\right) \qquad (5.25)$$

where a = βc, in which β is stress block parameter.

The predicted lateral strength of PT-MWs using the flexural expression is equal to the nominal moment capacity, M_n, divided by the effective height, h_n.

The method can be easily applied to design software as there is an iterative approach involved in the design procedure. The iteration process is required as c and f_{psi} are inter-related parameters in Eqs. 5.12 and 5.24. Note that the stresses in the PT bars located in the compression zone are less than the initial stresses.

Note that the expressions presented here were developed for purely unbonded PT-MWs. For hybrid walls, the effect of conventional reinforcement also needs to be considered.

5.9 Validation of the Proposed Design Approach

The strength predicted by the proposed design approach has been validated against experimental results as well as finite element model results and compared with the prediction considering no elongation of PT bars. MSJC (2013) has no procedure for estimating f_{ps} for unbonded PT-MWs. According to MSJC (2013), instead of a more accurate determination of f_{ps} for members with unbonded pre-stressing bars, it can conservatively be taken as f_{se}.

According to MSJC (2013), the base shear capacity of a PT-MW that does not incorporate bonded reinforcement, is the minimum strength obtained from the shear expression (Eq. 5.26) and the flexural expression.

$$V_n = min \begin{cases} 0.315A_n \sqrt{f'_m} & (a) \\ 2.07A_n & (b) \\ 0.621A_n + 0.45N & (c) \end{cases} \tag{5.26}$$

where A_n is the net cross sectional area of the wall.

The nominal moment capacity of a PT-MW using MSJC (2013) is:

$$M_n = \sum f_{se}A_{ps}\left(d - \frac{a}{2}\right) + N\left(\frac{L_w}{2} - \frac{a}{2}\right) \tag{5.27}$$

where: $a = \dfrac{\sum f_{se}A_{ps} + N}{0.8t_w f'_m}$

The predicted lateral strength of PT-MWs using the flexural expression is equal to the nominal moment capacity, M_n, divided by the effective height, h_n. As shown in Eq. 5.27, the MSJC (2013) does not take into account the distribution of PT bars or the stress increment due to the elongation of PT bars. However, the proposed expression (Eq. 5.25) considers both the elongation of the PT bars and their distribution along the length of the wall.

5.9.1 Validation of the Proposed Design Approach Against Experimental Results

To verify the accuracy of the proposed method compared with experimental results, following a comprehensive literature review of PT-MWs a database of 14 unbonded post-tensioned fully grouted specimens was collected. The summary of the selected walls is presented in Table 5.6.

Figure 5.15 compares the V_{EQN}/V_{EXP} calculated using the proposed method and the MSJC (2013) approach (No PT bar elongation) versus axial stress ratio. V_{EQN} is the minimum of the strength obtained from the shear equation (Eq. 5.26) and flexural expression. As shown, as the axial stress ratio increases, the prediction of MSJC (2013) becomes more conservative. However, the prediction from the

Table 5.6 Post-tensioned masonry wall database

References	Wall designation	Original designation	Specification	Material unit	Loading	h (mm)	t (mm)	L (mm)	Failure mode	Lateral strength (kN)
Laursen (2002)	L1-Wall1	FG:L3.0-W20-P3	FG	CMU	Cyclic	2800	190	3000	Flexural failure	560.9
	L1-Wall2	FG:L3.0-W15-P3	FG	CMU	Cyclic	2800	140	3000	Flexural failure	464.9
	L1-Wall3	FG: L3.0-W15-P2C	FG	CMU	Cyclic	2800	140	3000	Flexural failure	373.0
	L1-Wall4	FG: L3.0-W15-P2E	FG	CMU	Cyclic	2800	140	3000	Flexural failure	373.0
	L1-Wall5	FG:L1.8-W15-P2	FG	CMU	Cyclic	2800	140	1800	Flexural failure	177.9
	L1-Wall6	FG:L1.8-W15-P3	FG	CMU	Cyclic	2800	140	1800	Shear flexural failure	266.1
	L3- Wall1	S3-1	FG + Conf. plate	CMU	Cyclic	5250	140	2400	Flexural failure	212.0
	L3-Wall2	S3-2	FG + Conf. plate	CMU	Cyclic	5250	140	2400	Flexural failure	173.0
	L2-Wall1	FG: L3.0-W15-P1-CP	FG + Conf. plate	CMU	Cyclic	2800	140	3000	Flexural failure	249.0
	L2-Wall2	FG: L3.0-W15-P2-CP	FG + Conf. plate	CMU	Cyclic	2800	140	3000	Flexural failure, diagonal crack	396.0
	L2-Wall5	FG: L3.0-W15-P2-HB	FG + Conf. plate + high strength block	CMU	Cyclic	2800	140	3000	Flexural failure, diagonal crack	380.0
Rosenboom (2002))	R-Wall1	Test1	FG	Brick	Cyclic	2440	305	1220	Compression strut failure	330.9
	R-Wall2	Test3	FG + Conf. plate	Brick	Cyclic	2440	305	1220	Flexural failure	347.1
	R-Wall3	Test2	FG + suppl. mild steel	Brick	Cyclic	2440	305	1220	Flexural failure	365.6

FG fully grouted, *CMU* concrete masonry unit

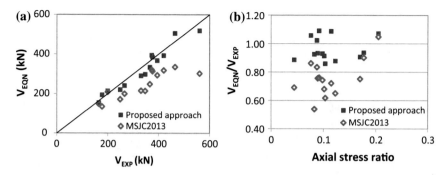

Fig. 5.15 **a** Comparison of the strength prediction and **b** comparison of V_{EQN}/V_{EXP}, using the proposed approach and MSJC 2013 approach based on experimental results

proposed method is unbiased toward the axial stress ratio and hence the proposed equation has effectively improved the strength prediction. While the value of V_{EQN}/V_{EXP} calculated according to MSJC (2013) varies from 0.54 to 1.05 with an average value of 0.75, it varies from 0.86 to 1.09 with an average of 0.96 using the proposed approach. Although for some tests the proposed approach over-predicts the lateral strength of the wall, the predicted strength falls within ±15% of the test results. Moreover, the proposed approach has reduced the scatter of the data compared with MSJC (2013).

5.9.2 Validation of the Proposed Design Approach Against Finite Element Results

The FEM results of walls set I and set II, were also used to investigate the accuracy of ignoring PT bar elongation in MSJC (2013) and compare it with the results obtained using the proposed approach. The base shear determined using the two wall sets are presented in Tables 5.7 and 5.8 and Fig. 5.16. The mode of failure and the ratio of V_{EQN}/V_{FEM} is also provided in Tables 5.7 and 5.8.

The value of V_{EQN}/V_{FEM} for walls that failed in flexure, calculated according to MSJC (2013), varies from 0.43 to 0.85 with an average value of 0.56 in set I, and varies from 0.49 to 0.69 with an average of 0.59 in set II. Using the proposed approach, V_{EQN}/V_{FEM} varies from 0.75 to 1.01 and 0.75 to 0.98, with an average of 0.91 and 0.88, for set I and set II, respectively. This comparison shows that while by ignoring the PT bar elongation the MSJC (2013) underestimates the flexural strength of PT-MWs, the proposed expression can effectively predict the strength. The MSJC (2013) approach failed in predicting the correct mode of failure. This occurs since the MSJC (2013) flexural prediction is too conservative, hence the predicted strength using the shear expression (Eq. 5.26) is greater than the predicted

Table 5.7 Strength prediction—set I of PT-MWs

Wall name	V_{FEM} (kN)	Failure mode	Proposed approach		MSJC 2013 (no PT bar elongation)			
					Flexural		Shear	
			V_{EQN} (kN)	V_{EQN}/V_{FEM}	V_{EQN} (kN)	V_{EQN}/V_{FEM}	V_{EQN} (kN)	V_{EQN}/V_{FEM}
W1-1	200	Flexure	198	0.99	106	0.53	446	2.23
W1-2	560	Flexure	500	0.89	280	0.5	614	1.10
W1-3*	714	Flexure	665	0.93	384	0.54	655	0.92
W1-4*	823	Shear	778	0.95	469	0.57	655	0.80
W1-5*	867	Shear	1202	1.39	737	0.85	655	0.76
W2-1	601	Flexure	526	0.87	280	0.47	614	1.02
W2-2	560	Flexure	503	0.90	280	0.5	614	1.10
W2-3	451	Flexure	407	0.90	280	0.62	614	1.36
W2-4	348	Flexure	327	0.94	276	0.79	610	1.75
W2-5*	644	Flexure	622	0.97	280	0.43	614	0.95
W2-6	560	Flexure	500	0.89	280	0.5	614	1.10
W2-7	400	Flexure	405	1.01	280	0.7	614	1.53
W2-8	330	Flexure	330	1.00	280	0.85	610	1.86
W3-1*	797	Shear	799	1.00	425	0.53	607	0.76
W3-2	560	Flexure	500	0.89	280	0.5	614	1.10
W3-3	386	Flexure	361	0.93	209	0.54	617	1.60
W3-4	300	Flexure	271	0.90	167	0.56	621	2.07
W4-1*	989	Shear	1573	1.59	699	0.71	975	0.99
W4-2*	1130	Shear	1538	1.36	748	0.66	1005	0.89
W4-3*	1200	Shear	1476	1.23	755	0.63	1009	0.84
W4-4*	955	Shear	1053	1.10	547	0.57	885	0.93
W4-5*	1090	Shear	1430	1.31	714	0.66	984	0.90
W5-1	193	Flexure	145	0.75	88	0.46	369	1.91
W5-2	560	Flexure	500	0.89	280	0.5	614	1.10
W5-3*	934	Shear	1076	1.15	543	0.58	852	0.91

*Lateral strength using proposed flexural expression is less than the shear expression of MSJC (2013)

strength using the flexure expression, implying that a flexural mode of failure will govern the response. For example, wall W4.2-3.8-5 failed in shear while according to the current approach in MSJC (2013) it is expected to exhibit a flexural mode of failure. Similar findings were made by Ryu et al. (2014). However, considering the proposed design approach, the predicted flexural strength was found to be higher than the predicted shear strength, calculated using Eq. 5.26, for the walls that failed in shear, indicating that the recommended design method could successfully predict the behavior and failure mode.

Table 5.8 Strength prediction—set II of PT-MWs

Wall name	V_{FEM} (kN)	Failure mode	Proposed approach		MSJC 2013 (no PT bar elongation)			
					Flexural		Shear	
			V_{EQN} (kN)	V_{EQN}/V_{FEM}	V_{EQN} (kN)	V_{EQN}/V_{FEM}	V_{EQN} (kN)	V_{EQN}/V_{FEM}
W1.8-2.8-1	138	Flexure	135	0.98	77	0.56	329	2.38
W1.8-2.8-2	200	Flexure	179	0.90	107	0.54	379	1.90
W1.8-2.8-3	257	Flexure	212	0.83	131	0.51	393	1.53
W1.8-2.8-4	297	Flexure	243	0.82	154	0.52	393	1.32
W1.8-2.8-5	331	Flexure	269	0.81	173	0.52	393	1.19
W1.8-3.8-1	100	Flexure	95	0.95	59	0.59	334	3.34
W1.8-3.8-2	166	Flexure	128	0.77	82	0.49	386	2.32
W1.8-3.8-3	186	Flexure	148	0.80	100	0.54	393	2.11
W1.8-3.8-4	200	Flexure	170	0.85	117	0.58	393	1.96
W1.8-3.8-5	215	Flexure	188	0.88	132	0.61	393	1.83
W1.8-4.8-1	78.8	Flexure	72	0.92	48	0.61	338	4.29
W1.8-4.8-2	126	Flexure	95	0.75	66	0.52	390	3.10
W1.8-4.8-3	145	Flexure	113	0.78	81	0.56	393	2.71
W1.8-4.8-4	149	Flexure	130	0.87	94	0.63	393	2.64
W1.8-4.8-5	164	Flexure	144	0.88	106	0.65	393	2.40
W3.0-2.8-1	389	Flexure	348	0.89	214	0.55	547	1.41
W3.0-2.8-2	538	Flexure	479	0.89	296	0.55	631	1.17
W3.0-2.8-3	647	Flexure	588	0.91	363	0.56	655	1.01
W3.0-2.8-4	727	Flexure	683	0.94	425	0.58	655	0.90
W3.0-2.8-5*	793	Shear	753	0.95	478	0.60	655	0.83
W3.0-3.8-1	293	Flexure	255	0.87	164	0.56	556	1.90
W3.0-3.8-2	400	Flexure	351	0.88	225	0.56	641	1.60
W3.0-3.8-3	486	Flexure	420	0.86	276	0.57	655	1.35
W3.0-3.8-4	520	Flexure	476	0.92	322	0.62	655	1.26
W3.0-3.8-5	568	Flexure	526	0.93	363	0.64	655	1.15
W3.0-4.8-1	218	Flexure	202	0.93	134	0.62	563	2.58
W3.0-4.8-2	320	Flexure	271	0.85	183	0.57	650	2.03
W3.0-4.8-3	366	Flexure	320	0.87	223	0.61	655	1.79
W3.0-4.8-4	395	Flexure	363	0.92	261	0.66	655	1.66
W3.0-4.8-5	430	Flexure	401	0.93	294	0.68	655	1.52
W4.2-2.8-1	767	Flexure	662	0.86	433	0.57	776	1.01
W4.2-2.8-2*	1014	Shear	925	0.91	590	0.58	890	0.88
W4.2-2.8-3*	1001	Shear	1129	1.13	706	0.70	917	0.92
W4.2-2.8-4*	1021	Shear	1316	1.29	843	0.83	917	0.90
W4.2-2.8-5*	1028	Shear	1481	1.44	942	0.92	917	0.89
W4.2-3.8-1	575	Flexure	492	0.86	331	0.58	788	1.37
W4.2-3.8-2	757	Flexure	679	0.90	450	0.59	906	1.20
W4.2-3.8-3	885	Flexure	818	0.92	539	0.61	917	1.04

(continued)

Table 5.8 (continued)

Wall name	V_{FEM} (kN)	Failure mode	Proposed approach		MSJC 2013 (no PT bar elongation)			
					Flexural		Shear	
			V_{EQN} (kN)	V_{EQN}/V_{FEM}	V_{EQN} (kN)	V_{EQN}/V_{FEM}	V_{EQN} (kN)	V_{EQN}/V_{FEM}
W4.2-3.8-4*	1010	Shear	955	0.95	643	0.64	917	0.91
W4.2-3.8-5*	1070	Shear	1054	0.99	723	0.68	917	0.86
W4.2-4.8-1	446	Flexure	394	0.88	272	0.61	800	1.79
W4.2-4.8-2	609	Flexure	533	0.87	367	0.60	917	1.51
W4.2-4.8-3	719	Flexure	642	0.89	438	0.61	917	1.28
W4.2-4.8-4	790	Flexure	734	0.93	524	0.66	917	1.16
W4.2-4.8-5	855	Flexure	806	0.94	588	0.69	917	1.07

*Lateral strength using proposed flexural expression is less than the shear expression of MSJC (2013)

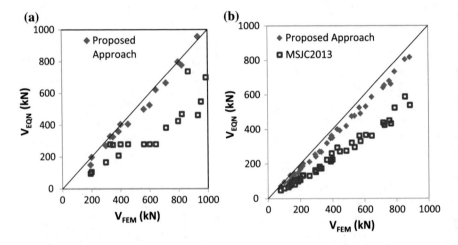

Fig. 5.16 Strength prediction of finite element models of **a** set I, and **b** set II

For the walls that failed in flexure, the flexural strength calculated using the proposed approach is compared with the MSJC (2013) approach, in Fig. 5.16a, b. While ignoring the bar elongation in MSJC (2013) yields a relatively conservative prediction, especially for the walls with higher lateral strength, the proposed design method provides a reasonable strength prediction.

In conclusion, comparing the strength and failure mode of the walls presented in Tables 5.7 and 5.8, with the results obtained from the proposed approach, reveals that considering the proposed approach for flexural strength, together with the current shear expression provided in MSJC (2013) (Eq. 5.26), yields an accurate prediction of the strength and failure mode.

5.10 Design Example

To provide a useful and designer-friendly approach to design an unbonded PT-MW, the proposed design procedure is presented in a flowchart (Fig. 5.17). An example design process is presented in this section to demonstrate the application of the proposed design method. The example wall has a height of 4 m, a length of 3 m, a thickness of 190 mm, is a concrete masonry wall, and is to be designed for a lateral strength of 300 kN. The yield strength and elastic modulus of the PT bars and the compressive strength of the masonry units are 1000 MPa, 200 GPa and 20 MPa, respectively.

To provide the maximum capacity of the wall the two PT bars are placed in the extreme cores (100 mm from the edge). To keep the spacing of PT bars as less than six times the nominal thickness of the wall, the third PT bar is considered to be at the central core providing 1000 mm PT bar spacing.

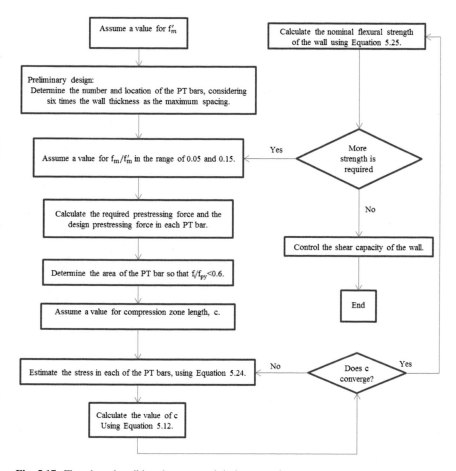

Fig. 5.17 Flowchart describing the proposed design procedure

To provide the maximum ductility and displacement capacity at peak strength and minimizing the applied axial stress, as the first trial, the value of f_m/f'_m is considered to be equal to 0.05, corresponding to an initial force of 139.3 kN in each PT bar. Assuming $f_i/f_{py} = 0.6$, the required steel area of each PT bar is 232.2 mm^2. Through iterative solution using Eqs. 5.12 and 5.24, the stress in the PT bars at peak strength was found to be equal to 1000, 972 and 532 MPa. Using Eq. 5.25, the moment capacity of the wall was determined to be 692.6 kN, corresponding to a base shear of 197.9 kN. The peak strength of the wall occurred at the wall top displacement of 30.3 mm corresponding to a drift ratio of 0.87%. As the lateral strength is not enough, the f_m/f'_m ratio can be increased; however as recommended it should be limited to 0.15. Considering this extreme limit and performing the same design process as explained for $f_m/f'_m = 0.05$, the initial force and the required area of each PT bar are found to be 418 kN and 696.7 mm^2, respectively. This results in stresses of 1000, 744 and 440 MPa in the PT bars at peak strength. The moment capacity and the corresponding base shear of the wall were determined to be 1729.5 kN m and 494.14 kN, respectively. The peak strength occurs at a wall top displacement of 21.35 mm which corresponds to a drift ratio of 0.61%. To reach the required lateral strength of 300 kN, a steel area of 379 mm^2 for each PT bar is required. This corresponds to a f_m/f'_m of 0.82, resulting in stresses of 1000, 861 and 504 MPa in the PT bars, and wall top displacement of 24.93 mm at the wall peak strength. Instead of lateral strength, lateral displacement at peak strength can be considered as design target.

This example illustrates that, as discussed previously, increasing the axial stress ratio increases the strength and reduces the displacement capacity and ductility. The shear capacity provided by MSJC (2013) (Eq. 5.26), for $f_m/f'_m = 0.5$ and 0.15 are found to be 447.6 and 588.8 kN, respectively. Both of these values are higher than the values calculated for flexural strength, hence, a shear failure mode is not expected to occur. The two extreme ratios of f_m/f'_m considered in the example cover the strength of 692.6 kN to 1729.5 kN m and rotation at peak strength of 0.61–0.87%. For a desired displacement and/or strength within these ranges, a value of $0.05 \leq f_m/f'_m \leq 0.15$ can be considered. For design requirements beyond these limits, a designer may consider changing other parameters (e.g. the locations and numbers of PT bars, the wall thickness and masonry compressive strength).

In another study (Hassanli et al. 2014c) the accuracy of the proposed approach is compared with the existing expressions including the equations provided in masonry codes and other existing approaches, using available test results and finite element model results.

5.11 Conclusion

This chapter uses finite element models to develop a design expression to predict the in-plane flexural strength of post-tensioned masonry walls (PT-MWs). To calibrate the material model, a finite element model of a masonry prism was developed and calibrated with experimental results. The numerical models of six large scale PT-MWs were developed and validated against experimental results. A parametric study was performed to investigate the effect of different parameters on the behavior and strength of PT-MWs. Multivariate regression analysis was performed to develop an equation to evaluate the rotation of PT-MWs at the peak strength. The developed equation was incorporated into the flexural analysis of PT-MWs. The flexural and shear strength of 14 wall specimens collected from the literature and the FEM walls were calculated using the MSJC (2013) and the proposed approach. This showed that the proposed approach leads to accurate and rational evaluation of the flexural strength of unbonded PT-MWs. It is concluded that the proposed expression could significantly improve the strength prediction of PT-MWs and that disregarding the elongation of the PT bars in the unbonded PT-MW results in a highly conservative strength prediction. Moreover, the predicted mode of failure might not be accurate due to low predicted flexural strength.

Based on the results of this study, the following design recommendations are proposed:

- The axial stress ratio should be kept to not greater than 0.15. For axial stress ratios over 0.15, increasing the applied PT force slightly increases the strength. However, it has been shown by Hassanli et al. (2014b) that increasing the axial stress ratio beyond 0.15 leads to a brittle mode of failure.
- The initial stress in the PT bars is recommended to be less than 60% of the yield strength of the bars.
- An iterative solution using Eqs. 5.12 and 5.24 is recommended to estimate the stress developed in the PT bars. (Note that the accuracy of the proposed equation for parameters beyond the limits provided in this study, needs to be investigated).
- Although in this manuscript equations are proposed for unbonded PT-MWs, preliminary investigation by the authors revealed that the recommended equation could accurately predict the strength of post-tensioned concrete walls and can be applied to these members as well.

Notation

The following symbols are used in this chapter. The definitions for other symbols are presented here in the manuscript.

ε_0	Masonry compressive strain at decompression point	M_n	Nominal moment capacity of the wall
Δ_0	Wall displacement at decompression point	t_w	Wall thickness
Δf_{psi}	The stress increment in a PT bar due to elongation Δ_i	θ_0	Wall rotation at decompression point
Δ_i	Elongation in PT bar i	θ_m	Wall rotation at peak strength
A_n	Net cross sectional area of the wall	φ_0	Maximum curvature at decompression point

(continued)

(continued)

ε_0	Masonry compressive strain at decompression point	M_n	Nominal moment capacity of the wall
$A_{ps\,i}$	Area of PT bar i	a	Depth of an equivalent compression stress block at nominal strength
d_i	Distance from the extreme compression fibre to the ith PT bar	c	Compression zone length at peak strength
E_m	Elastic modulus of masonry	d	Distance from extreme compression fiber to centroid of tension reinforcement
E_{ps}	Young's modulus of the pre-stressing steel	f_m	Axial stress
f'_m	Compressive strength of masonry	f_{py}	Yield strength of PT bar
$f_{se\,i}$	Effective stress in the ith PT bar after stress losses	N	gravity load
h_w	Wall height	α, β	Stress block parameters
L_{ps}	Unbonded length of the PT bar	θ	Wall rotation
L_w	Wall length	υ	Poisson's ratio

References

Bean Popehn JR, Schultz AE, Drake CR (2007) Behavior of slender, post-tensioned masonry walls under transverse loading. J Struct Eng 133(11):1541–1550

Crawford J, Malvar L (1997) User's and theoretical manual for K&C concrete model. Karagozian & Case, Burbank, CA, TR-97-53.1

ElGawady MA, Sha'lan A (2011) Seismic behavior of self-centering precast segmental bridge bents. J Bridge Eng ASCE 16(3):328–339

ElGawady MA, Booker AJ, Dawood H (2010) Seismic behavior of post-tensioned concrete-filled fiber tubes. J Compos Const ASCE 14(5):616–628

Hassanli R, Elgawady MA, Mills JE (2014a) An evaluation of design code expressions for estimating in-plane shear strength of partially grouted masonry walls. Aust J Struct Eng 15(3):299–316

Hassanli R, Elgawady MA, Mills JE (2014b) Strength and seismic performance factors of post-tensioned masonry walls. J Struct Eng (in press)

Hassanli R, Elgawady MA, Mills JE (2014c) Simplified approach to predict the flexural strength of unbonded post-tensioned masonry walls. J Struct Eng (under review)

Henry RS, Sritharan S, Ingham JM (2012b) Unbonded tendon stresses in post-tensioned concrete walls at nominal flexural strength. ACI Struct J 109(2)

Laursen PPT (2002) Seismic analysis and design of post-tensioned concrete masonry walls. Ph.D. dissertation, Department of Civil and Environmental Engineering, University of Auckland, Auckland, New Zealand

Lourenço PB (1996) A matrix formulation for the elastoplastic homogenisation of layered materials. Mech Cohesive-Frictional Mater 1(3):273–294

LS-DYNA Keyword User's Manual (2007) Livermore Software Technology, California, USA

Magallanes JM, Wu Y, Malvar LJ, Crawford JE (2010) Recent improvements to release III of the K&C concrete model. In: Proceedings of 11th international LS-DYNA Users conference, 6–8 June. Dearborn, MI

Malvar LJ, Crawford JE, Wesevich JW, Simons D (1994) A new concrete material model for DYNA3D. TM-94-14.3. Report to the Defense Nuclear Agency. Glendale, Karagozian and Case Structural Engineers, CA

Malvar LJ, Crawford JE, Wesevich JW, Simons D (1997) A plasticity concrete material model for DYNA3D. Int J Impact Eng 19(9–10):847–873

Malvar LJ, Crawford JE, Wesevich JW, Simons D (1996) A new concrete material model for DYNA3D release II: shear dilation and directional rate enhancement. TR-96-2.2, report to the Defense Nuclear Agency. Glendale, Karagozian and Case Structural Engineers, CA

Malvar LJ, Crawford JE, Morill KB (2000) K&C concrete material model release III: automated generation of material model input. TR-99-24.3, technical report, Karagozian and Case Structural Engineers, Glendale, CA

Masonry Standards Joint Committee (MSJC) (2013) Building code requirements for masonry structures. ACI 530/ASCE 5, TMS 402, American Concrete Institute, Detroit

Noble C (2007) DYNA3D finite element analysis of steam explosion loads on a pedestal wall design. Lawrence Livermore National Laboratory (LLNL), Livermore, California, USA

Nolph S, ElGawady MA (2012) Static cyclic response of partially grouted masonry shear walls. J Struct Eng 138(7):864–879

Priestley M, Elder D (1983) Stress-strain curves for unconfined and confined concrete masonry. Proc ACI J Proc ACI 80(3):192–201

Rosenboom OA (2002) Post-tensioned clay brick masonry walls for modular housing in seismic regions. MS. thesis, North Carolina State University, Raleigh, NC, USA

Ryu D, Wijeyewickrema A, ElGawady M, Madurapperuma M (2014) Effects of tendon spacing on in-plane behavior of post-tensioned masonry walls. J Struct Eng 140(4), CID:04013096

Scott BD, Park R, Priestley MJN (1982) Stress-strain behavior of concrete confined by overlapping hoops at low and high strain rates. ACI J 79(1):13–27

Wight GD, Kowalsky MJ, Ingham JM (2007) Direct displacement-based seismic design of unbonded post-tensioned masonry walls. ACI Struct J 104(5):560–569

Chapter 6
Simplified Approach to Predict the Flexural Strength of Unbonded Post-tensioned Masonry Walls

A simplified design approach is developed in this chapter to predict the flexural strength of unbonded PT-MWs. The accuracy of different flexural expressions is also investigated in this chapter according to experimental and finite element modelling results. An analytical procedure is developed to predict the force displacement response of PT-MWs. The accuracy of the analytical model is then validated against available experimental test results for unconfined and confined PT-MWs. Using a similar analytical procedure, a parametric study is performed to obtain the force-displacement response of walls with different features. Multivariate regression analysis is performed to develop an empirical equation to estimate the compression zone length in unbonded PT-MWs. The proposed equation for compression zone length is then incorporated into the flexural analysis of post-tensioned masonry walls and validated against experimental results and finite element results.

6.1 Introduction

Recent research has demonstrated that unbonded post-tensioned structural elements including concrete walls, concrete columns, and masonry walls can display high ductility levels while withstanding high levels of seismic loads. When a slender unbonded masonry wall (PT-MW) is subjected to a lateral in-plane load, usually a single horizontal crack forms at the wall-foundation interface. However, in squat unbounded PT-MWs, the failure can be characterized by inclined cracks (shear or flexural-shear cracks) or vertical cracks (due to high compressive stresses in the toe)

A modified version of this chapter has been published in the Journal of Engineering Structures Hassanli R., ElGawady M. A. and Mills J. E., Simplified approach to predict the flexural strength of unbonded post-tensioned masonry walls, Journal of Engineering Structures, Volume 142, PP 255–271, 2017.

© Springer International Publishing AG, part of Springer Nature 2019
R. Hassanli, *Behavior of Unbounded Post-tensioned Masonry Walls*,
Springer Theses, https://doi.org/10.1007/978-3-319-93788-5_6

instead. In a PT-MW with a rocking response, the restoring nature of the post-tensioning (PT) force returns the wall to its original vertical position and minimizes the residual displacement. This behavior is specifically favorable for structures that are designed for immediate occupancy performance levels. The rocking mechanism of PT-MWs results in plastic deformation concentrated at the toe of the wall, which can be repaired with minimal cost (Wight 2006; Bean Popehn et al. 2007; ElGawady et al. 2010; ElGawady and Sha'lan 2011; Dawood et al. 2012; Ryu et al. 2014; Hassanli et al. 2014a).

To determine the in-plane flexural strength of an unbonded PT-MW, the level of stress developed in PT bars corresponding to the wall peak strength needs to be calculated. The stress developed in a PT bar is a function of the bar strain and hence the elongation of the bars. In bonded PT-MWs, the strain compatibility concept can be considered to determine the stress in the bars. For unbonded PT-MWs, the strain in the PT bar remains approximately constant along the length of the bar. Therefore, instead of the conventional strain compatibility equations used for strain calculations in structural elements having bonded reinforcement, displacement compatibility criteria need to be considered, in which the stress in the PT bars is a function of wall rotation and neutral axis depth. The current approach of the Masonry Standards Joint Committee (MSJC 2013) ignores the stress increase in PT bars beyond initial post-tensioning. However, several experimental and finite element studies have shown that under lateral loads the post-tensioning force increased (Laursen 2002; Wight 2006; Bean Popehn et al. 2007; Ryu et al. 2014). Expressions have been proposed by different researchers for evaluating such post-tensioning force increases under in-plane loading (Wight 2006) and out-of-plane loading (Bean Popehn et al. 2007).

The primary objectives of the research presented in this chapter are:

- To compare the accuracy of different expressions in predicting the in-plane flexural strength of unbonded PT-MWs based on experimental and finite element model results.
- To elaborate on an existing procedure to obtain the lateral force-displacement response of unbonded PT-MWs.
- To perform a parametric study to develop an empirical equation to estimate the compression zone length and a non-iterative expression to estimate the flexural strength of unbonded PT-MWs.

6.2 Prediction of Nominal Flexural Strength

This section reviews different expressions available in the literature to predict the flexural strength of PT-MWs. These expressions include MSJC (2013) and methods A, B and C based on published research.

6.2.1 Masonry Standard Joint Committee (MSJC 2013)

MSJC (2013) ignore the elongation of bars in the PT-MWs, hence, the following equations can be considered to predict the flexural strength of PT-MWs:

$$M_n = \left(f_{se}A_{ps} + f_yA_s + N\right)\left(d - \frac{a}{2}\right) \tag{6.1}$$

$$a = \frac{f_{se}A_{ps} + f_yA_s + N}{0.8f'_m b} \tag{6.2}$$

where a is the depth of the equivalent compression zone, A_s is the area of conventional flexural reinforcement, f_y is the yield strength, f_{se} is the effective stress in the PT bar after immediate stress losses, A_{ps} is the area of the PT bar, N is the gravity load including the self-weight of the wall, f'_m is the compressive strength of masonry, b is the cross section width and d is the effective depth of the wall. The predicted lateral strength of PT-MWs using this flexural expression is equal to the nominal moment capacity, M_n, divided by the effective height, h_n.

The shear capacity, according to MSJC (2013), of PT-MWs having no bonded steel can be calculated as follows:

$$V_n = min \begin{cases} 0.315A_n\sqrt{f'_m} & (a) \\ 2.07A_n & (b) \\ 0.621A_n + 0.45N & (c) \end{cases} \tag{6.3}$$

where A_n is the net cross sectional area of the wall.

In the flexural expression presented by MSJC (2013), different locations of PT bars are not considered. The equation was originally developed for out-of-plane loading in which the PT bars are usually located at the center of the wall, resulting in a single value of d. While acceptable for out-of-plane loading, for in-plane loading the equation is not able to account for the distribution of multiple PT bars along the length of the wall. Hence, Eqs. 6.1 and 6.2 need to be re-written as follows:

$$M_n = \sum f_{ps\,i}A_{ps\,i}\left(d_i - \frac{a}{2}\right) + \sum f_yA_{s\,i} + N\left(\frac{L_w}{2} - \frac{a}{2}\right) \tag{6.4}$$

$$a = \frac{\sum f_{ps\,i}A_{ps\,i} + \sum f_yA_{s\,i} + N}{0.8f'_m b} \tag{6.5}$$

where L_w is the length of the wall. The predicted lateral strength of PT-MWs using the flexural expression is equal to the nominal moment capacity, M_n, divided by the effective height, h_n.

For unbonded PT-MWs under in-plane loading the MSJC (2013) uses Eq. 6.6 to evaluate f_{ps}.

$$f_{ps} = f_{se} \quad \text{(In-plane bending)} \tag{6.6}$$

It is worth noting that Eq. 6.6 does not take into account the stress increment due to the elongation of PT bars.

For unbonded PT-MWs under out-of-plane bending, Eq. 6.7 is considered by the MSJC (2013) to evaluate f_{ps},

$$f_{ps} = f_{se} + 0.03 \left(\frac{E_{ps}d}{L_p} \right) \left(1 - 1.56 \frac{A_{ps}f_{ps} + N}{f'_m L_w d} \right) \quad \text{(Out-of-plane bending)} \tag{6.7}$$

where L_p is the unbonded length and E_{ps} is the elastic modulus of PT bar.

Method A: Bean Popehn, Schultz and Drake's Approach

Ryu et al. (2014) indicated that the out-of-plane expression of MSJC (2008) provides a reasonable estimate of flexural strength for walls loaded in-plane. The equation was proposed by Bean Popehn et al. (2007) as a result of a series of laboratory tests and finite element models of PT-MWs loaded out-of-plane. However, the equation has been updated in the latest version of MSJC (2013) (Eq. 6.7). Moreover, to determine the in-plane flexural strength of PT-MWs having multiple post-tensioning bars, the ultimate stress in each PT bar needs to be calculated. Hence, Eq. 6.7 can be re-written as follows:

$$f_{ps\,i} = f_{se} + 0.03 \left(\frac{E_{ps}d_i}{L_p} \right) \left(1 - 1.56 \frac{\sum A_{ps\,i} f_{ps\,i} + N}{f'_m L_w d_i} \right) \tag{6.8}$$

Equation 6.8 can be solved iteratively for $f_{ps\,i}$.

Note that Eq. 6.7 was developed by Bean Popehn et al. (2007) for wall loaded out-of-plan, with a symmetric section and a single point-tensioning tendon at mid-depth, and Eq. 6.8 is inspired form Eq. 6.7, to investigate its potential to be considered for in-plane loading as well.

Method B: Wight and Ingham's Approach

Equation 6.8 assumed a constant rotation of PT-MWs of 0.03 rad (or drift of 0.03). However, it has been reported that the rotation of walls at the peak strength is not constant and is a function of the configuration of the wall, aspect ratio, and axial stress ratio. Using experimental results and finite element models, Wight and Ingham (2008) proposed Eq. 6.9 to estimate the peak tendon force:

$$f_{ps} = f_{se} + \frac{E_{ps}}{l_p} \theta \left(d_i - \frac{f_m L_w}{\alpha \beta f'_m} \right) \tag{6.9}$$

where

$$f_m = \frac{f_{se}A_{ps} + N}{L_w t_w} \tag{6.10}$$

$$\theta = \frac{\left(\frac{h_w}{L_w}\right)\varepsilon_{mu}}{30\left(\frac{f_m}{f'_m}\right)} \tag{6.11}$$

where α and β are the stress block parameters and ε_{mu} is the ultimate masonry strain, which are provided by different building codes (e.g. in MSJC (2013): $\alpha = = 0.8$, $\varepsilon_{mu} = 0.0035$ and 0.0025 for clay and concrete masonry, respectively).

Method C: Hassanli et al.'s Approach

Hassanli et al. (2014b) developed Eq. 6.12 to predict the tendon ultimate stress of PT-MWs,

$$f_{ps\,i} = f_{se\,i} + (\theta_m c - \theta_0 c)\frac{E_{ps}}{L_p}\left(\frac{d_i}{c} - 1\right) \tag{6.12}$$

where,

$$\theta_m c = (0.00055L_w + 17.375\frac{f_m}{f'_m}) \tag{6.13}$$

$$\theta_0 = \left(\frac{1}{1350}\right)\frac{f_m}{f'_m}\frac{h_w}{L_w} \quad \text{Concrete masonry} \tag{6.14}$$

$$\theta_0 = \left(\frac{1}{900}\right)\frac{f_m}{f'_m}\frac{h_w}{L_w} \quad \text{Clay masonry} \tag{6.15}$$

$$c = \frac{\sum f_{ps\,i}A_{ps\,i} + N}{\alpha\beta f'_m t_w} \tag{6.16}$$

where c is the length of the compression zone, θ_m is the wall rotation at peak strength and θ_0 is the rotation corresponding to the decompression point.

Both the compression zone length c, and $f_{ps\,i}$ are unknown but can be determined by simultaneously solving Eqs. 6.12 and 6.16.

In the following section, data collected from literature based on experimental and finite element results are used to investigate the accuracy of the presented expressions.

6.3   Comparison of Flexural Expressions with Experimental Results

Table 6.1 summarizes a database of 14 unbonded fully grouted PT-MWs tested under in-plane loading. The walls had heights ranging from 2800 to 5250 mm, lengths ranging from 1000 to 3000 mm, compressive strengths ranging from 13.3 to 20.6 MPa and axial stress ratios ranging from 0.04 to 0.2. The axial stress ratio is defined as f'_m/f_m, where f_m is defined using Eq. 6.10.

The minimum strength obtained from the shear strength equation (Eq. 6.3) according to the MSJC (2013) and each of the three flexural strength approaches for each wall is calculated as V_{EQN}. V_{EXP} is the maximum strength reported during the experimental work. The values of V_{EQN}/V_{EXP} are presented in Table 6.2.

Figure 6.1 presents the relationship between V_{EQN} and V_{EXP} based on the results in Table 6.2. As shown in the figure and table, by ignoring the elongation of PT bars, the MSJC (2013) underestimated the strength of 93% of the test specimens.

Table 6.1 Post-tensioned masonry wall database

Reference	Wall designation	Original designation	Specification	material unit	Loading
Laursen (2002)	L1-Wall1	FG:L3.0-W20-P3	FG	CMU	Cyclic
	L1-Wall2	FG:L3.0-W15-P3	FG	CMU	Cyclic
	L1-Wall3	FG: L3.0-W15-P2C	FG	CMU	Cyclic
	L1-Wall4	FG: L3.0-W15-P2E	FG	CMU	Cyclic
	L1-Wall5	FG:L1.8-W15-P2	FG	CMU	Cyclic
	L1-Wall6	FG:L1.8-W15-P3	FG	CMU	Cyclic
	L3-Wall1	S3-1	FG + confinement plate	CMU	Cyclic
	L3-Wall2	S3-2	FG + confinement plate	CMU	Cyclic
	L2-Wall1	FG: L3.0-W15-P1-CP	FG + confinement plate	CMU	Cyclic
	L2-Wall2	FG: L3.0-W15-P2-CP	FG + confinement plate	CMU	Cyclic
	L2-Wall5	FG: L3.0-W15-P2-HB	FG + confinement plate + high strength block	CMU	Cyclic
Rosenboom (2002)	R-Wall1	Test1	FG	Brick	Cyclic
	R-Wall2	Test3	FG + confinement plate	Brick	Cyclic
	R-Wall3	Test2	FG + supplemental mild steel	Brick	Cyclic

FG fully grouted, *CMU* concrete masonry unit

Table 6.2   Strength prediction using different approaches

Wall	V_{EQN}/V_{EXP}			
	MSJC (2013) (no PT bar elongation)	Method A	Method B	Method C
L1-Wall1	0.54	1.00	0.63	0.93
L1-Wall2	0.72	1.18	0.79	1.09
L1-Wall3	0.86	1.06	0.93	1.06
L1-Wall4	0.76	0.94	0.85	0.93
L1-Wall5	0.76	1.17	0.84	1.09
L1-Wall6	0.75	1.01	0.77	0.91
L3- Wall1	0.65	0.91	0.71	0.88
L3-Wall2	1.05	1.14	1.05	1.07
L2-Wall1	0.90	1.02	0.91	0.94
L2-Wall2	0.62	0.88	0.68	0.86
L2-Wall5	0.68	0.93	0.73	0.91
R-Wall1	0.69	0.89	0.79	0.89
R-Wall2	0.74	0.96	0.79	0.93
R-Wall3	0.83	1.04	0.89	1.02
Max	1.05	1.18	1.05	1.09
Min	0.54	0.88	0.63	0.86
Average	0.75	1.01	0.81	0.96
Std dev.	0.13	0.10	0.11	0.08
Var.	0.02	0.01	0.01	0.01
Range	0.51	0.30	0.42	0.23

According to Table 6.2, the value of V_{EQN}/V_{EXP} based on MSJC (2013) varies from 0.54 to 1.05 with a range of 0.51, an average of 0.75, and standard deviation of 0.13. *Method A* predicts V_{EQN}/V_{EXP} values varying from 0.88 to 1.18 with a range of 0.30, an average of 1.01, and standard deviation of 0.10. However, *Method A* over predicted the strength of 50% of the test specimens. Using *Method B*, V_{EQN}/V_{EXP} varies from 0.63 to 1.05 with a range of 0.42, an average of 0.81, and standard deviation of 0.11. Similar to MSJC (2013), *Method B* underestimated the strength of 93% of the test specimens. However, it has a better average and narrower range compared to the MSJC (2013) results. *Method C* presents the lowest range and most accurate conservative average of V_{EQN}/V_{EXP}. Using *Method C*, V_{EQN}/V_{EXP} varies from 0.86 to 1.09 with a range of 0.23, and an average of 0.96. *Method C* has the lowest standard deviation of 0.08 among the four approaches. However, *Method C* over-predicted the strength of 36% of the test specimens. Adopting a strength reduction factor of 0.9, all approaches except *Method A* underestimated the strength of the test specimens. *Method A* still over predicted the strength of 21% of the test specimens.

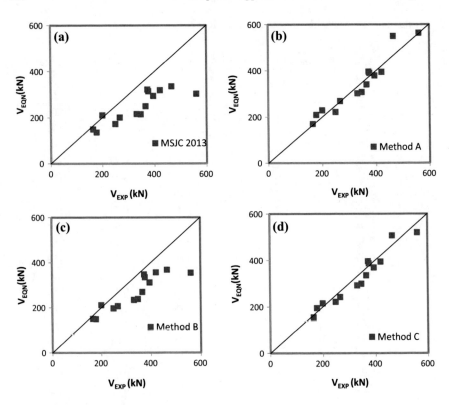

Fig. 6.1 Accuracy of different expressions in predicting the base shear—based on experimental results

Figure 6.2 presents the values of V_{EQN}/V_{EXP} versus the axial stress ratio. As shown in the figure, the MSJC (2013) approach, *Method A, Method B, and Method C* have R^2 values of 0.34, 0.12, 0.19, and 0.02, respectively, indicating that *Method C* is the least biased toward the level of axial stress ratio compared to the other approaches.

6.4   Comparison of Flexural Expressions with Finite Element Model Results

Finite element models of PT-MWs were developed by Hassanli et al. (2014b). The models were validated against experimental results of six PT-MWs and were used to carry out a parametric study on 45 walls referenced as set II in Hassanli et al. (2014b). This set of data is considered in this manuscript to investigate the ability of different expressions to predict the flexural strength. Within this database, 40 specimens that displayed a flexural failure were considered. The walls in the set had

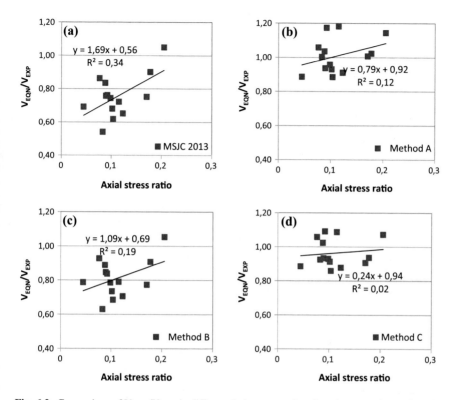

Fig. 6.2 Comparison of V_{EQN}/V_{FEM} in different design expressions based on experimental results

axial stress ratios ranging from 0.05 to 0.15, heights ranging from 2800 to 4800 mm, and lengths ranging from 1800 to 4200 mm.

Figure 6.3 presents the predicted base shear obtained from the finite element analysis, V_{FEM}, versus the predicted base shear obtained using the different approaches. Figure 6.4 presents the values of V_{EQN}/V_{FEM} obtained using the different approaches. As shown in the figures, the MSJC (2013) approach is the most conservative approach. This is expected since the MSJC (2013) ignores the post-tensioning bar elongation. For *Method A*, while the FEM results follow the experimental results quite well, it overestimated the strength of 44% of the investigated walls with a standard deviation of 0.129 and an average of V_{EQN}/V_{FEM} of 1.02. Moreover, the approach tends to be more un-conservative as the applied axial load ratio increases. *Method B* provides a relatively over- conservative estimation for all specimens. *Method C* followed very closely the FEM results with an average of V_{EQN}/V_{FEM} of 0.91 and standard deviation of 0.06. *Method C* overestimated the strength of 7% of the FE walls, however, considering a strength reduction factor of 0.9, it provides a conservative estimate of the strength of all the walls.

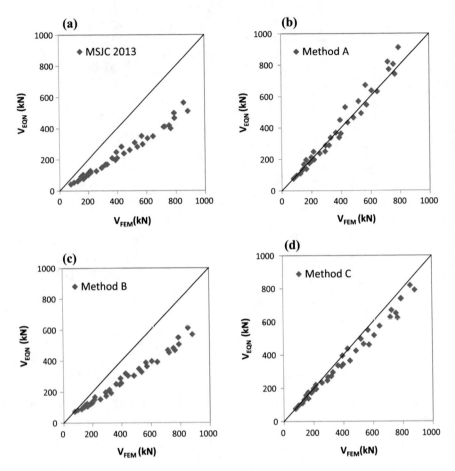

Fig. 6.3 Comparison of V_{EQN}/V_{FEM} using the different design expressions based on FEM results

6.5 Proposed Simplified Method

While the PT-MW strength prediction method proposed by Hassanli et al. (2014b) provides relatively better results compared with the other examined methods, it includes an iterative procedure, which may become cumbersome and time-consuming for walls with multiple PT bars. Design spreadsheets would be required to implement the approach. In this manuscript, a simplified method is developed to avoid the iteration involved in Hassanli et al.'s approach.

To develop the simplified method, an existing analytical approach was improved and used to predict the force-displacement responses of a series of walls. Then, a multi-variate regression analysis was performed to develop an empirical equation to estimate the compression zone length, c. This equation was then integrated into the expression proposed by Hassanli et al. (2014b) to provide a simplified method in which iteration is avoided.

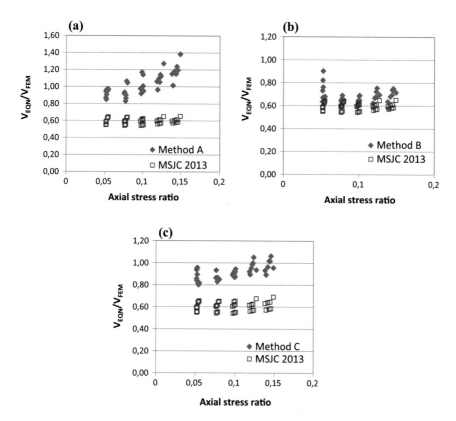

Fig. 6.4 Comparison of ignoring PT bar elongation in MSJC (2013) with other expressions in predicting the base shear—based on FEM results

6.5.1 Stress in Unbonded PT Bars

Formation of inelastic deformations usually occurs within the plastic zone. Instead of the actual physical length over which the plastic deformation spreads (L_{pz}), an equivalent plastic hinge length, L_{pl} can be considered within which the plastic curvature is assumed to be lumped, and the remaining height of the wall is assumed to remain linearly elastic (Paulay and Priestly 2009). L_{pl} is smaller than L_{pz} and can be used as an approximate approach to determine the displacement capacity of a wall. The concept of equivalent plastic hinge length, considering a constant ultimate curvature over a specific plastic hinge length, has been developed mainly to facilitate the seismic design of a structure. Considering the plastic deformation lumped within the plastic hinge length and ignoring the rotation due to elastic deformations outside the plastic hinge zone, the rotation of a masonry wall can be determined as (Paulay and Priestly 2009):

$$\theta = \int_0^{L_{pl}} \frac{\varepsilon_{mu}}{c} y dy = \frac{\varepsilon_{mu}}{c} L_{pl} \qquad (6.17)$$

where ε_{mu} is the ultimate masonry strain. The other components are defined in Fig. 6.5a.

In an unbonded cantilever wall with a flexural mode of failure and a rocking mechanism (Fig. 6.5a), the wall's rotation can be expressed as:

$$\theta = \frac{\Delta_i}{d_i - c} \qquad (6.18)$$

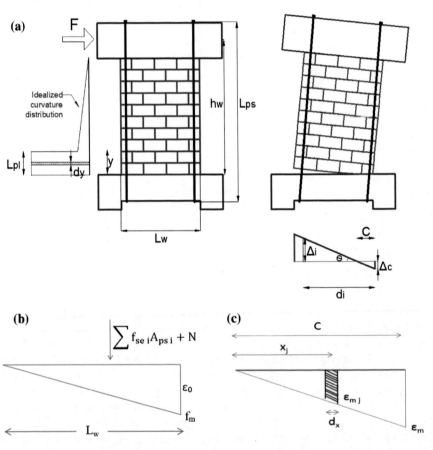

Fig. 6.5 PT-MW **a** before and after deformation, **b** stress distribution at decompression of the heel, **c** strain of a typical element in the compression zone

Therefore:

$$\theta_m = \frac{\Delta_i}{d_i - c} = \frac{\varepsilon_{mu}}{c} L_{pl} \qquad (6.19)$$

Hence, according to Fig. 6.5a the elongation in PT bar i, is equal to:

$$\Delta_i = \frac{\varepsilon_{mu}}{c} L_{pl}(d_i - c) \qquad (6.20)$$

and the stress increment due to elongation Δ_i is:

$$\Delta f_{ps\,i} = \frac{\varepsilon_{mu} E_{ps} L_{pl}}{c L_{ps}} (d_i - c) \qquad (6.21)$$

Hence, the total stress developed in the ith PT bar can be determined as:

$$f_{ps\,i} = f_{se\,i} + L_{pl}(\varepsilon_{mu} - \varepsilon_0) \frac{E_{ps}}{L_{ps}} \left(\frac{d_i}{c} - 1 \right) \qquad (6.22)$$

where $f_{se\,i}$ is the effective stress in the ith PT bar after stress losses, L_{ps} is the unbonded length of the PT bar, E_{ps} is the Young's modulus of the pre-stressing steel, c is the compression zone length, and d_i is the distance from the extreme compression fibre to the ith PT bar. By neglecting the mortar tensile strength, the rocking of the wall starts after the wall experiences stresses higher than the decompression stress in the heel. Considering a linear stress-strain relationship in masonry, assuming plane sections remain plane and ignoring the elongation of PT bars before the decompression point, according to Fig. 6.5b, the absolute maximum masonry compressive strain corresponding to the decompression point is:

$$\varepsilon_0 = \frac{2(\sum f_{se\,i} A_{ps\,i} + N)}{l_w t_w E_m} \qquad (6.23)$$

where A_{ps} is the area of the PT bar, E_m is the elastic modulus of masonry, t_w is the thickness of the wall, and N is the gravity load. Considering a maximum value of 0.15 for f_m/f'_m, as recommended by Hassanli et al. (2014a), limits the stress in the masonry corresponding to the decompression point to $0.3 f'_m$. Hence, considering a linear stress-strain relationship in the masonry at the decompression point is a reasonable assumption.

The strain in the toe region is the summation of the strain before and after the decompression point. According to MSJC (2013), the elastic moduli of concrete masonry and clay masonry can be considered as $900 f'_m$, and $700 f'_m$ respectively The axial stress, f_m, is defined as:

$$f_m = \frac{\sum f_{se\,i} A_{ps\,i} + N}{L_w t_w} \tag{6.24}$$

Hence,

$$\varepsilon_0 = \left(\frac{1}{450}\right) \frac{f_m}{f'_m} \quad \text{Concrete masonry} \tag{6.25}$$

$$\varepsilon_0 = \left(\frac{1}{300}\right) \frac{f_m}{f'_m} \quad \text{Clay} \tag{6.26}$$

where f'_m is the compressive strength of masonry.

Using equilibrium, the compression zone length, c, can be expressed as:

$$c = \frac{\sum f_{ps\,i} A_{ps\,i} + N}{\alpha \beta f'_m t} \tag{6.27}$$

where α and β are the stress block parameters which are provided by different building codes (e.g. in MSJC 2013: $\alpha = \beta = 0.8$).

In order to determine the stress in the tendon using Eq. 6.22, the values of the two critical parameters ε_{mu} and L_{pl}, need to be determined.

6.5.2 Ultimate Masonry Strain, ε_{mu}

It has been reported that an ultimate strain of 0.003 is a valid assumption for conventional reinforced concrete and bonded PT concrete (Harajli 2006). However, Henry et al. (2012) and Dawood et al. (2011) found that this value resulted in a conservative estimate of the rocking strength of unconfined concrete walls. The accuracy of the value of 0.003 for ultimate compressive strain of concrete is more questionable for unbonded walls, as the maximum strain, ε_{cu}, at the onset of crushing may deviate significantly, depending on the loading pattern and distribution of cracks. According to MSJC (2013), the ultimate masonry strain is 0.0035 and 0.0025 for clay and concrete masonry, respectively. The average masonry compressive strain reported by Shedid et al. (2010) based on an experimental study on masonry walls was found to be at least 0.005 for most of the tested walls. However, for members with high moment gradients, the maximum strain can reach 0.006–0.008 (Paulay and Priestly 2009). The values of ε_{cu} for bonded walls are not appropriate for unbonded walls, mainly due to the rapid change in the curvature near the wall base. As a result of the steep change in the strain at the toe region, experimentally, it is difficult to accurately measure the concrete compression strain at the extreme fiber in an unbonded PT-MW (Henry et al. 2012). The strain recorded using different instrumentation is scattered significantly as a result of localized failure at the toe of the wall. Using finite element models, Henry et al.

(2012) found that a strain limit of 0.005 may be more suitable for describing the flexural strength of PT concrete walls. Hence, a value of 0.005 was assumed in this study as the ultimate masonry strain.

6.5.3 Plastic Hinge Length

In conventional reinforced concrete columns, because of the bond between reinforcement and concrete, the strain, cracks and plastic deformation seem to be distributed over a higher length of the column compared to columns having unbonded PT bars (ElGawady et al. 2010). For members having bonded reinforcement, curvatures, rotations, and displacements can be determined by considering equilibrium and strain compatibility. However, to do so, a plastic hinge length is required. The plastic hinge length in a concrete/masonry wall depends on different wall parameters including: dimensions, material properties, the area and spacing of vertical and horizontal steel, support conditions and the type and magnitude of loading, bond slip between grout and reinforcement and strain yield penetration (Mortezaei and Ronagh 2012). Because of the complexity of the problem, simplified experimental expressions have been proposed to evaluate the plastic hinge length. Paulay and Priestley (1992) and Priestley et al. (2007) proposed Eqs. 6.28–6.30, respectively, to evaluate the plastic hinge length of concrete walls:

$$L_{pl} = 0.2L_w + 0.044h_w \tag{6.28}$$

$$L_{pl} = 0.08L_w + 0.022f_y d_b \tag{6.29}$$

$$L_{pl} = 0.2\left(\frac{f_{pu}}{f_{py}} - 1\right)h_w + 0.1L_w + 0.022f_{py}d_b \tag{6.30}$$

where L_w = wall length, h_w = wall height, f_{py}, f_{pu} and d_b are the yield strength, ultimate strength and bar diameter of the longitudinal reinforcement.

In a more recent study, using finite element analysis, Bohl and Adebar (2011) considered the effect of axial compression in the evaluation of the plastic hinge length of concrete shear walls (Eq. 6.31):

$$L_{pl} = (0.2L_w + 0.05l_v)\left(1 - 1.5\frac{P}{A_w f'_m}\right) \leq 0.8L_w \tag{6.31}$$

where P = axial force, and A_w = wall area.

Instead of the equivalent plastic hinge length concept, an analytical model was developed by Shedid and El-Dakhakhni (2014) to estimate the displacements of reinforced masonry structural walls.

In the British masonry code BS 5628-2 (2005), L_{pl} is assumed to be equal to c. Substituting L_{pl} with c in Eq. 6.19:

$$\theta_m = \frac{\varepsilon_{mu}}{c} c \rightarrow \theta_m = \varepsilon_{mu} \tag{6.32}$$

The ε_{mu}-value is a code defined constant, hence for all walls the British masonry code considers the same rotational capacity regardless of the configuration of the wall and level of post-tensioning force.

To measure the plastic hinge length experimentally, the strain profile needs to be recorded at different heights of the wall. In an unbonded wall at the toe region the strain changes rapidly within a relatively small length, which is difficult to measure. Consequently, the plastic hinge length is a function of the type, location and length of the measurement devices (Henry et al. 2012). Hence, other methods such as analytical approaches, which are considered in this study, must be used to evaluate the plastic hinge length. Substituting Eq. 6.13 proposed by Hassanli et al. (2014b) in Eq. 6.19 and considering a value of 0.005 for ε_{mu}, the plastic hinge length, L_{pl} can be defined as:

$$L_{pl} = 0.11 L_w + 3475 \frac{f_m}{f'_m} \tag{6.33}$$

Assuming that the strain is significant within a length equal to the plastic hinge length for any value of wall rotation, θ, Eq. 6.22 can be rewritten as:

$$f_{ps\,i} = f_{se\,i} + L_{pl}(\varepsilon_m - \varepsilon_0) \frac{E_{ps}}{L_{ps}} \left(\frac{d_i}{c} - 1 \right) \tag{6.34}$$

In the following section, Eqs. 6.33 and 6.34 are used to predict the force-displacement behavior of PT-MWs.

6.6   Analytical Procedure to Obtain Force-Displacement Response of PT-MWs

An analytical approach originally proposed by Rosenboom and Kowalsky (2004) to predict the force-displacement response of unbonded PT-MWs has been used and improved. The modified procedure considers the plastic hinge length expression proposed by Hassanli et al. (2014b) and the stress-strain relationships proposed by Priestley and Elder (1983), developed for confined and unconfined masonry based on a series of prism tests.

6.6.1   Decompression Point

PT bars display an increase in their initial post-tensioning stresses when the wall-footing interface joint opens. Before this opening, the PT force remains constant. Moreover, under the initial post-tensioning force, stresses in the masonry are usually less than 15% f'_m (Hassanli et al. 2014a). Considering a linear stress-strain relationship in the masonry, and assuming that plane sections remain plane, the absolute maximum masonry compressive strain corresponding to the decompression point at the bottom-most point of the PT-MW is (Fig. 6.5b):

$$\varepsilon_0 = \frac{2\left(\sum f_{se\,i} A_{ps\,i} + N\right)}{L_w t_w E_m} \tag{6.35}$$

where E_m is the elastic modulus of masonry = $750f'_m$ and $900f'_m$ for clay and concrete masonry, respectively (MSJC 2013).

The base shear V can be determined using Eq. 6.36:

$$V_0 = \frac{\sum f_{se\,i} A_{ps\,i} d_i + N\left(\frac{L_w}{2}\right) - c_{m0}\left(\frac{L_w}{3}\right)}{h_w} \tag{6.36}$$

where

$$c_{m0} = 0.5 L_w t_w \varepsilon_0 E_m \tag{6.37}$$

The lateral displacement at the top of the wall corresponding to the decompression state is:

$$\Delta_0 = \frac{\varphi_0 h_w^2}{3} \tag{6.38}$$

where φ_0 is the maximum value of the curvature at the decompression point $= \varepsilon_0 / L_w$.

6.6.2   Force-Displacement Response Beyond Decompression State

To determine the remaining portion of the backbone curve i.e. lateral displacement versus lateral force, an iterative procedure is used as shown below:

Step 1: Assuming top displacement, Δ, calculate the rotation of the wall, $\theta = \Delta/h_w$

Step 2: Assume a value of compression zone length, c

Step 3: Calculate the strain in each PT bar, $\varepsilon_{ps\,i}$, and the masonry strain at the toe, ε_m using Eqs. 6.39 and 6.40, which were derived based on displacement compatibility and assuming constant curvature over the plastic hinge length.

$$\varepsilon_{ps\,i} = \theta(d_i - c)/L_{ps} + \varepsilon_{i\,initial} \tag{6.39}$$

$$\varepsilon_m = \theta c/L_{pl} + \varepsilon_0 \tag{6.40}$$

where $\varepsilon_{i\,initial}$ is the initial strain in the ith PT bar after immediate losses and L_{pl} can be determined using Eq. 6.33.

Step 4: Calculate the stress developed in each PT bar using Eq. 6.41 and assuming elasto-plastic stress-strain relationships (Priestley et al. 2007):

$$\sigma_{ps\,i} = \begin{cases} \varepsilon_{ps\,i}E_{ps} & \varepsilon_{ps\,i} \le \varepsilon_{yp} \\ \varepsilon_{yp}E_{ps} + \left(\varepsilon_{ps\,i} - \varepsilon_{yp}\right)\left(f_{up} - f_{yp}\right)/\left(\varepsilon_{ps\,i} - \varepsilon_{yp}\right) & \varepsilon_{yp} < \varepsilon_{ps\,i} \le \varepsilon_{up} \\ 0 & \varepsilon_{up} < \varepsilon_{ps\,i} \end{cases} \tag{6.41}$$

where ε_{yp} and ε_{up}, are the yield and ultimate strain, and f_{yp} and f_{up} are the corresponding yield and ultimate stresses, respectively.

Step 5: Calculate the corresponding masonry stress using the modified Kent-Park stress-strain relationships proposed by Priestley and Elder (1983) (Eq. 6.42):

$$f_{mj}(\varepsilon_{mj}) = \begin{cases} 1.067kf'_m\left[\left(\frac{2\varepsilon_{mj}}{0.002}\right) - \left(\frac{\varepsilon_{mj}}{0.002}\right)^2\right] & \varepsilon_{mj} < 0.0015 \\ kf'_m\left[1 - Z_m\left(\varepsilon_{mj} - 0.0015\right)\right] & 0.0015 \le \varepsilon_{mj} \le \varepsilon_{mp} \\ 0.2k'_m & \varepsilon_{mj} > \varepsilon_{mp} \end{cases} \tag{6.42}$$

where

$$Z_m = \frac{0.5}{\left[\frac{3+0.29f'_m}{145f'_m - 1000}\right] + \frac{3}{4}\rho_s\sqrt{\frac{h''}{s_h}} - 0.002K} \tag{6.43}$$

$$\varepsilon_{mp} = \frac{0.8}{Z_m} + 0.0015 \tag{6.44}$$

$$K = 1 + \rho_s\frac{f_{yh}}{f'_m} \tag{6.45}$$

$$\varepsilon_{mj} = (x_i/c)\varepsilon_m \tag{6.46}$$

where ρ_s and f_{yh} are the volumetric ratio and the confining steel yield strength, h'' is the lateral dimension of the confined core and s_h is the longitudinal spacing of confining steel. For walls without a confining plate, $\rho_s = 0$ and $K = 1$. ε_{mj} is the masonry strain at distance x_j as shown in Fig. 6.5c.

Step 6: Calculate the total compression force, c_m, and the total tension force, T:

$$C_m = \int_0^c f_{mj}dA = \int_0^c f_{mj}l_w dx \tag{6.47}$$

$$T = \sum T_i = \sum \sigma_{ps\,i}A_{ps\,i} \tag{6.48}$$

where σ_{psi} and f_{mj} can be determined using Eqs. 6.41 and 6.42, respectively.

Step 7: If $c_m = T$ go to the next step, otherwise go back to step 3.

Step 8: Calculate $\Delta_f = \Delta_0 + \Delta$

Step 9: Take the moment about the neutral axis to calculate the total moment capacity, M:

$$M = \sum_{i=1}^n T_i(d_i - c) + N(0.5L_w - c) - \int_0^c f_{mj}x dA \tag{6.49}$$

where n is the number of PT bars.

Step 10: The terms $(\Delta_f, V = M/h_w)$ correspond to a point in the force-displacement curve. To obtain another point, go to step 2. Figure 6.6 summarizes the iterative procedure.

6.7 Validation of the Analytical Procedure

To illustrate the capability and accuracy of the described analytical method, eleven fully grouted PT-MWs were selected from the literature and analyzed, consisting of six walls without confinement and five walls with a confinement plate. The configurations of these walls are presented in Table 6.3.

All walls were concrete masonry and were tested under in-plane increasing cyclic displacement until failure occurred (Laursen 2002). In all walls except L3-1 and L3-2, the PT bars had yield strength, tensile strength, and elastic modulus of 970 MPa, 1160 MPa and 190 GPa, respectively while for walls L3-1 and L3-2 the PT strands had yield strength, tensile strength, and elastic modulus of 1520 MPa, 1785 MPa and 190 GPa, respectively (Laursen 2002). An ultimate strain of 0.08 was used for the strands and bars.

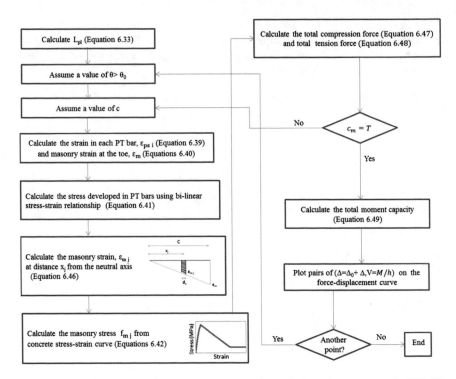

Fig. 6.6 Analytical procedure flowchart to obtain the force-displacement response of a PT-MWs

The lateral force-displacement results obtained from the analytical models are compared with the backbone curves obtained from the experimental cyclic tests in Figs. 6.7 and 6.8 for walls with and without confinement plates, respectively. As shown in the figures, for both confined and unconfined walls, the model can correctly predict the wall strength, initial stiffness and rotational capacity. The model can also capture the post-peak response accurately. Table 6.4 compares the strength of each wall obtained from the experimental results, V_{EXP}, with the strength obtained from the analytical approach, $V_{analysis}$. According to the table, the predicted strength of the specimens using the analytical approach falls within $\pm 10\%$ of the average of the test results in the pull and push directions. Figures 6.9 and 6.10 compare the toughness (energy absorption per cycle) versus displacement curves for the walls with and without confinement plates, respectively. As shown, the analytical approach can effectively predict the wall's toughness for different displacement values. The toughness obtained from the analytical approach for different displacement values lower than the ultimate displacement, falls within $\pm 12\%$ of the corresponding experimental results.

Table 6.3 Configuration of the selected experimental walls

Wall designation	Original designation	Specification	h (mm)	t (mm)	L (mm)	f'_m (MPa)	PT bar initial stress (f_{se}) (MPa)	Initial PT force on the wall (kN)
L1-Wall2	FG:L3.0-W15-P3	FG	2800	140	3000	15.1	555	690
L1-Wall3	FG:L3.0-W15-P2C	FG	2800	140	3000	20.6	757	622
L1-Wall4	FG:L3.0-W15-P2E	FG	2800	140	3000	17.8	757	628
L1-Wall5	FG:L1.8-W15-P2	FG	2800	140	1800	20.5	534	445
L1-Wall6	FG:L1.8-W15-P3	FG	2800	140	1800	18.4	614	760
L2-Wall1	FG:L3.0-W15-Pl-CP	FG + confinement plate	2800	140	3000	18.2	720	299
L2-Wall2	FG: L3.0-W15-P2-CP	FG + confinement plate	2800	140	3000	15.1	703	584
L2-Wall5	FGrL3.0-Wls-P2-IIB	FG + confinement plate + high strength block	2800	140	3000	17.8	743	617
L3-Wall1	S3-1	FG + confinement plate	5250	140	2400	17.9	948	395.1
L3-Wall2	S3-2	FG + confinement plate	5250	140	2400	14	1017	423.9

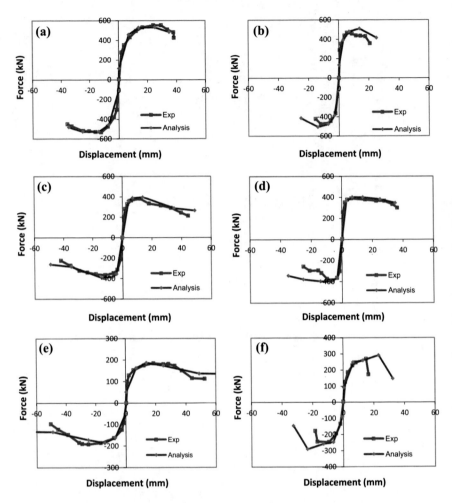

Fig. 6.7 Force displacement curves of walls **a** L1-Wall1, **b** L1-Wall2, **c** L1-Wall3, **d** L1-Wall4, **e** L1-Wall5 and **f** L1-Wall6

6.8 Proposed Expression for Compression Zone Length

Using the verified analytical approach, a parametric study was carried out for a hypothetical set of PT-MWs to investigate the effects of different factors on the compression zone length c. In total, 90 walls were compiled and analyzed. The configurations of the walls were developed according to the recommendations provided by Hassanli et al. (2014a), i.e. the axial stress ratio varied from 0.05 to 0.15%, the f_i/f_{py} value ranged from 0.3 to 0.6 and the spacing between PT bars was kept lower than six times the wall thickness. For all walls $f'_m = 13.3$ MPa, and the yield strength and elastic modulus of 970 MPa and 190 GPa, respectively were

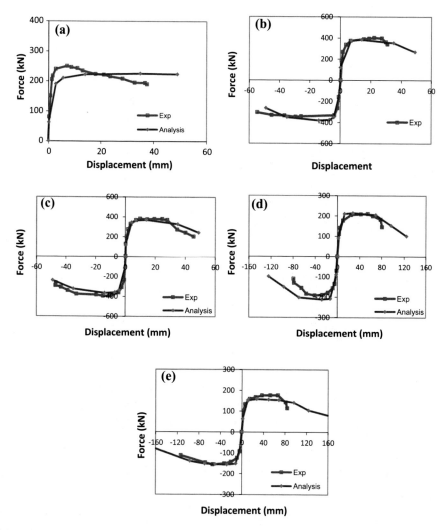

Fig. 6.8 Force displacement curves of walls **a** L2-Wall1, **b** L2-Wall2, **c** L2-Wall5, **d** L3-Wall1 and **e** L3-Wall2

adopted for the PT bars. Table 6.5 presents the matrix of the considered walls. As shown in the table, three different lengths of 1800, 3000 and 4200 mm, three heights of 2800, 3800 and 4800 mm, two thicknesses of 90 and 190 mm, and five different axial stress ratios ranging from 0.05 to 0.15, were considered to establish the matrix.

For the walls presented in Table 6.5, force-displacement curves were developed using the described analytical approach. Using the analytical models, the compression zone length, c, corresponding to the maximum lateral strength was

Table 6.4 Prediction of the strength using analytical approach

Wall	$V_{analysis}$ (kN)	V_{EXP} (kN)			$V_{analysis}/V_{EXP}$		
		Pull	Push	Average	Pull	Push	Average
L1-Wall1	529.0	535.6	550.2	542.9	0.99	0.96	0.97
L1-Wall2	508.0	464.4	477.6	471.0	1.09	1.06	1.08
L1-Wall3	394.9	378.2	368.1	373.2	1.04	1.07	1.06
L1-Wall4	399.2	386.2	390.3	388.2	1.03	1.02	1.03
L1-Wall5	186.8	183.4	193.2	188.3	1.02	0.97	0.99
L1-Wall6	284.8	248.5	267.1	257.8	1.15	1.07	1.10
L2-Wall1	225.0	249.5		249.5	0.90		0.90
L2-Wall2	382.4	349.1	400.1	374.6	1.10	0.96	1.02
L2-Wall5	367.6	397.8	378.9	388.4	0.92	0.97	0.95
L3-Wall1	217.7	192.8	208.4	200.6	1.13	1.04	1.09
L3-Wall2	157.0	156.9	175.8	166.4	1.00	0.89	0.94

obtained for all 90 walls (Table 6.6). As shown in Table 6.6, the compression zone length at the peak strength, ranged from 120 to 1175 mm, corresponding to 6.7 and 28.0% of L_w, respectively.

Different factors influenced the compressive zone length including: length of the wall, axial stress ratio and reinforcement ratio. Figure 6.11 presents the effect of the height, length, thickness of the wall and axial stress ratio on the compression zone length. The upward trend of the data and high slope of the regression line presented in Figs. 6.11b, d show that as the wall length and axial stress ratio increases the compression zone length increases. In contrast, the small value of R^2 and small slope of the regression lines in Figs. 6.11a, c, imply that the compression zone length is effectively independent of the thickness and height of the wall.

Using the compression zone length obtained for the dataset (Table 6.6), a multivariate regression analysis was carried out. As a result, the wall's length and the axial pre-stress ratio were found to be the most influential factors affecting the compression zone length. Considering these two parameters, the following equation was developed from the multi-variate regression analysis to estimate the compression zone length in unbonded PT-MWs:

$$c = \alpha L_w \frac{f_m}{f'_m} \tag{6.50}$$

The R^2-value and adjusted R^2-value obtained from the regression analysis were 99 and 98%, respectively, implying that the proposed equation provides an acceptable prediction of the compression zone length, c. The adjusted R^2 in a multi-variate regression analysis is a modified version of R^2 that has been adjusted for the number of predictors in the model. The value of α obtained from the curve fit was 1.7, however for design purposes a value of 2.0 is considered to be appropriate. The application of this expression is limited to wall members having a flexural mode of failure.

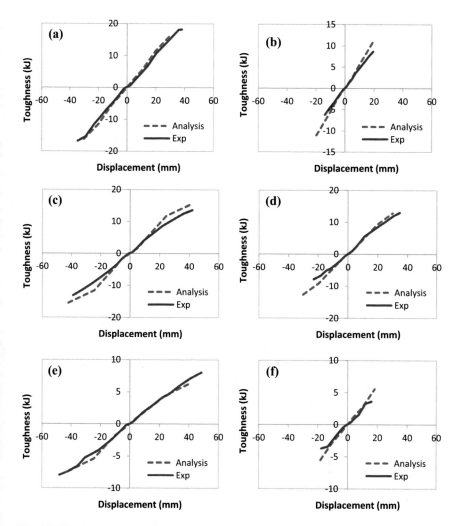

Fig. 6.9 Toughness versus displacement of walls **a** L1-Wall1, **b** L1-Wall2, **c** L1-Wall3, **d** L1-Wall4, **e** L1-Wall5 and **f** L1-Wall6

In Eq. 6.12 the value of θ_0 is considerably smaller than θ_m, and hence can be ignored (Hassanli et al. 2014b). Substituting Eq. 6.50 in Eq. 6.12 and ignoring θ_0, the following simplified expression is proposed to predict the stress developed in the ith PT bar in an unbonded PT-MW:

$$f_{ps\,i} = f_{se\,i} + \left(0.00055 L_w + 17.375 \frac{f_m}{f_m'}\right) \frac{E_{ps}}{L_p} \left(\frac{d_i f_m'}{2 L_w f_m} - 1\right) \tag{6.51}$$

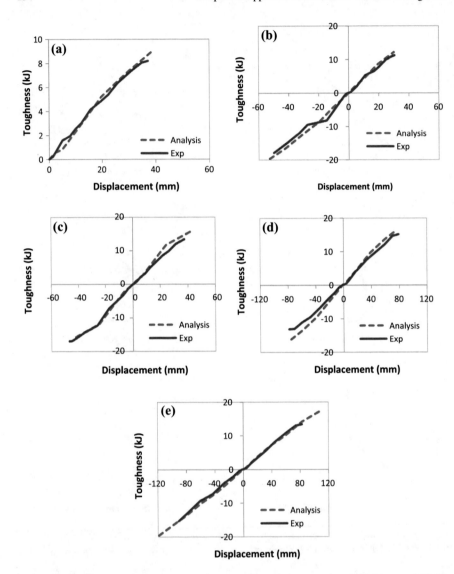

Fig. 6.10 Toughness versus displacement of walls **a** L2-Wall1, **b** L2-Wall2, **c** L2-Wall5, **d** L3-Wall1 and **e** L3-Wall2

6.9 Validation of the Proposed Simplified Approach

The strengths predicted by the proposed simplified approach were then validated against experimental results and finite element model results, and also compared with the predicted values obtained considering no PT bar elongation which is allowed by MSJC (2013).

Table 6.5 Wall matrix of parametric study

Wall	t_w (m)	h (m)	L_w (m)	f_m/f'_m	No. of tendons	Initial stress, f_{se} (MPa)
W1	0.19	2.8	1.8	0.052	2	482.3
W2	0.19	2.8	1.8	0.077	2	467.9
W3	0.19	2.8	1.8	0.098	2	455.8
W4	0.19	2.8	1.8	0.12	2	444.7
W5	0.19	2.8	1.8	0.14	2	433.3
W6	0.19	3.8	1.8	0.052	2	486.4
W7	0.19	3.8	1.8	0.078	2	476.0
W8	0.19	3.8	1.8	0.1	2	463.9
W9	0.19	3.8	1.8	0.122	2	454.6
W10	0.19	3.8	1.8	0.144	2	445.5
W11	0.19	4.8	1.8	0.053	2	489.5
W12	0.19	4.8	1.8	0.078	2	478.7
W13	0.19	4.8	1.8	0.101	2	470.1
W14	0.19	4.8	1.8	0.124	2	461.1
W15	0.19	4.8	1.8	0.146	2	452.3
W16	0.19	2.8	3.0	0.052	3	533.4
W17	0.19	2.8	3.0	0.076	3	517.6
W18	0.19	2.8	3.0	0.097	3	502.8
W19	0.19	2.8	3.0	0.119	3	490.5
W20	0.19	2.8	3.0	0.139	3	476.9
W21	0.19	3.8	3.0	0.052	3	539.5
W22	0.19	3.8	3.0	0.077	3	524.9
W23	0.19	3.8	3.0	0.099	3	513.0

Wall	t_w (m)	h (m)	L_w (m)	f_m/f'_m	No. of tendons	Initial stress, f_{se} (MPa)
W46	0.09	2.8	1.8	0.052	2	482.3
W47	0.09	2.8	1.8	0.077	2	467.9
W48	0.09	2.8	1.8	0.098	2	455.8
W49	0.09	2.8	1.8	0.12	2	444.7
W50	0.09	2.8	1.8	0.14	2	433.3
W51	0.09	3.8	1.8	0.052	2	486.4
W52	0.09	3.8	1.8	0.078	2	476.0
W53	0.09	3.8	1.8	0.1	2	463.9
W54	0.09	3.8	1.8	0.122	2	454.6
W55	0.09	3.8	1.8	0.144	2	445.5
W56	0.09	4.8	1.8	0.053	2	489.5
W57	0.09	4.8	1.8	0.078	2	478.7
W58	0.09	4.8	1.8	0.101	2	470.1
W59	0.09	4.8	1.8	0.124	2	461.1
W60	0.09	4.8	1.8	0.146	2	452.3
W61	0.09	2.8	3.0	0.052	3	533.4
W62	0.09	2.8	3.0	0.076	3	517.6
W63	0.09	2.8	3.0	0.097	3	502.8
W64	0.09	2.8	3.0	0.119	3	490.5
W65	0.09	2.8	3.0	0.139	3	476.9
W66	0.09	3.8	3.0	0.052	3	539.5
W67	0.09	3.8	3.0	0.077	3	524.9
W68	0.09	3.8	3.0	0.099	3	513.0

(continued)

Table 6.5 (continued)

Wall	t_w (m)	h (m)	L_w (m)	f_m/f'_m	No. of tendons	Initial stress, f_{se} (MPa)
W24	0.19	3.8	3.0	0.121	3	500.3
W25	0.19	3.8	3.0	0.142	3	489.1
W26	0.19	4.8	3.0	0.053	3	543.6
W27	0.19	4.8	3.0	0.078	3	530.3
W28	0.19	4.8	3.0	0.101	3	519.5
W29	0.19	4.8	3.0	0.123	3	508.8
W30	0.19	4.8	3.0	0.145	3	498.7
W31	0.19	2.8	4.2	0.054	4	582.4
W32	0.19	2.8	4.2	0.078	4	553.9
W33	0.19	2.8	4.2	0.096	4	523.2
W34	0.19	2.8	4.2	0.121	4	523.2
W35	0.19	2.8	4.2	0.139	4	504.1
W36	0.19	3.8	4.2	0.054	4	588.6
W37	0.19	3.8	4.2	0.079	4	564.7
W38	0.19	3.8	4.2	0.099	4	537.5
W39	0.19	3.8	4.2	0.124	4	539.6
W40	0.19	3.8	4.2	0.145	4	525.9
W41	0.19	4.8	4.2	0.055	4	596.8
W42	0.19	4.8	4.2	0.08	4	572.8
W43	0.19	4.8	4.2	0.101	4	545.7
W44	0.19	4.8	4.2	0.127	4	552.7
W45	0.19	4.8	4.2	0.149	4	539.5
W69	0.09	3.8	3.0	0.121	3	500.3
W70	0.09	3.8	3.0	0.142	3	489.1
W71	0.09	4.8	3.0	0.053	3	543.6
W72	0.09	4.8	3.0	0.078	3	530.3
W73	0.09	4.8	3.0	0.101	3	519.5
W74	0.09	4.8	3.0	0.123	3	508.8
W75	0.09	4.8	3.0	0.145	3	498.7
W76	0.09	2.8	4.2	0.054	4	582.4
W77	0.09	2.8	4.2	0.078	4	553.9
W78	0.09	2.8	4.2	0.096	4	523.2
W79	0.09	2.8	4.2	0.121	4	523.2
W80	0.09	2.8	4.2	0.139	4	504.1
W81	0.09	3.8	4.2	0.054	4	588.6
W82	0.09	3.8	4.2	0.079	4	564.7
W83	0.09	3.8	4.2	0.099	4	537.5
W84	0.09	3.8	4.2	0.124	4	539.6
W85	0.09	3.8	4.2	0.145	4	525.9
W86	0.09	4.8	4.2	0.055	4	596.8
W87	0.09	4.8	4.2	0.08	4	572.8
W88	0.09	4.8	4.2	0.101	4	545.7
W89	0.09	4.8	4.2	0.127	4	552.7
W90	0.09	4.8	4.2	0.149	4	539.5

Table 6.6 Compression zone length, c, to wall length, L_w

Wall	c (m)	c/L_w	Wall	c (m)	c/L_w	Wall	c (m)	c/L_w	Wall	c (m)	c/L_w
W1	0.19	0.11	W24	0.645	0.22	W46	0.19	0.11	W69	0.645	0.22
W2	0.28	0.16	W25	0.74	0.25	W47	0.28	0.16	W70	0.74	0.25
W3	0.365	0.20	W26	0.25	0.08	W48	0.25	0.14	W71	0.31	0.10
W4	0.32	0.18	W27	0.435	0.15	W49	0.32	0.18	W72	0.455	0.15
W5	0.395	0.22	W28	0.654	0.22	W50	0.395	0.22	W73	0.655	0.22
W6	0.19	0.11	W29	0.78	0.26	W51	0.19	0.11	W74	0.625	0.21
W7	0.185	0.10	W30	0.723	0.24	W52	0.185	0.10	W75	0.725	0.24
W8	0.25	0.14	W31	0.365	0.09	W53	0.25	0.14	W76	0.395	0.09
W9	0.32	0.18	W32	0.59	0.14	W54	0.32	0.18	W77	0.48	0.11
W10	0.39	0.22	W33	0.77	0.18	W55	0.39	0.22	W78	0.77	0.18
W11	0.12	0.07	W34	0.93	0.22	W56	0.125	0.07	W79	0.93	0.22
W12	0.2	0.11	W35	1.06	0.25	W57	0.2	0.11	W80	1.06	0.25
W13	0.245	0.14	W36	0.365	0.09	W58	0.25	0.14	W81	0.365	0.09
W14	0.315	0.18	W37	0.585	0.14	W59	0.315	0.18	W82	0.585	0.14
W15	0.38	0.21	W38	0.82	0.20	W60	0.38	0.21	W83	0.82	0.20
W16	0.255	0.09	W39	0.99	0.24	W61	0.255	0.09	W84	0.927	0.22
W17	0.44	0.15	W40	1.065	0.25	W62	0.495	0.17	W85	1.065	0.25
W18	0.545	0.18	W41	0.443	0.11	W63	0.545	0.18	W86	0.365	0.09
W19	0.655	0.22	W42	0.58	0.14	W64	0.655	0.22	W87	0.585	0.14
W20	0.76	0.25	W43	0.795	0.19	W65	0.76	0.25	W88	0.795	0.19
W21	0.315	0.11	W44	0.92	0.22	W66	0.255	0.09	W89	1	0.24
W22	0.413	0.14	W45	1.175	0.28	W67	0.435	0.15	W90	1.175	0.28
W23	0.54	0.18				W68	0.54	0.18			

6.9.1 Validation of the Proposed Simplified Design Approach Against Experimental Results

Experimental results presented in Table 6.1 were used to verify the accuracy of the proposed method. Figure 6.12 compares the V_{EQN}/V_{EXP} value calculated using the proposed approach and the no elongation approach (MSJC 2013) versus the axial stress ratio. V_{EQN} is the minimum strength obtained from the shear equation (Eq. 6.3) and flexural expression. As shown, as the axial stress ratio increases the prediction of MSJC (2013) becomes more conservative; however, the prediction from the proposed simplified method is relatively unbiased toward the axial stress ratio. The proposed equation has therefore effectively improved the strength prediction. While the value of V_{EQN}/V_{EXP} calculated using MSJC (2013) varied from 0.54 to 1.05 with an average of 0.75, it ranged from 0.86 to 1.09 with an average of 0.96 using the proposed simplified approach. The predicted strength obtained from the proposed approach falls within ±15% of the test results. Moreover, as shown in Fig. 6.12, the proposed approach has reduced the scatter of the data compared with MSJC (2013).

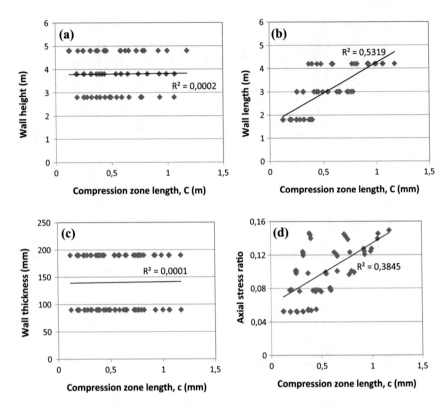

Fig. 6.11 Effect of **a** wall length, **b** wall height, **c** wall thickness and **d** axial stress ratio on the compression zone length

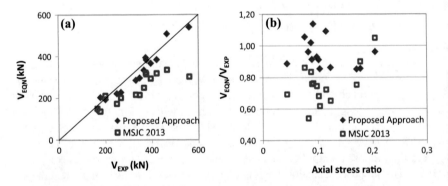

Fig. 6.12 **a** Comparison of the strength prediction and **b** comparison of V_{EQN}/V_{EXP} using the proposed approach and MSJC (2013) approach—according to experimental result

6.9.2   Validation of the Proposed Simplified Design Approach Against Finite Element Results

The FEM results of the set I and set II of walls developed by Hassanli et al. (2014b) were used to compare the prediction of MSJC (2013) and the proposed simplified approach. The wall configuration and detail of the finite element models can be found in Chap. 5.

The strengths of the walls that exhibited a flexural failure, calculated using the proposed approach and compared with the MSJC (2013) approach are illustrated in Figs. 6.13a, b, for Set I and Set II, respectively. By ignoring the elongation of PT bars, the MSJC (2013) approach yields a relatively conservative prediction, especially for the walls having higher lateral strength, whereas the proposed design method provides a more reasonable strength prediction.

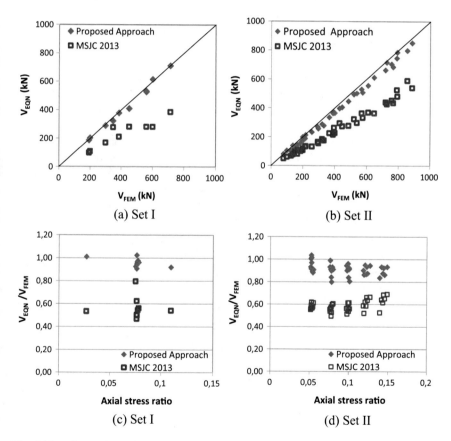

Fig. 6.13 **a** Comparison of the strength prediction and **b** comparison of $V_{EQN}/V_{EXP,}$ using the proposed approach and MSJC (2013)

As shown in Figs. 6.13c, d, the value of V_{EQN}/V_{FEM} for the walls in set I which failed in flexure calculated according to MSJC (2013) varied from 0.46 to 0.79 with an average value of 0.54, and varied from 0.49 to 0.69 with an average of 0.59 for set II. Using the proposed approach, V_{EQN}/V_{FEM} varies from 0.78 to 0.99 and 0.8 to 1.03, with an average of 0.90 and 0.91, for set I and set II, respectively. This comparison confirms that while the MSJC (2013) underestimates the flexural strength of PT-MWs, the proposed expression can effectively improve the prediction.

6.10 Conclusion

This chapter compares the accuracy of different expressions in predicting the flexural strength of PT-MWs. A new expression for wall rotation and a constitutive model were implemented in an existing analytical procedure to predict the force-displacement response of PT-MWs. The procedure was validated against experimental test results of 11 PT-MWs. The proposed procedure correctly predicted the force-displacement response, lateral strength, stiffness and post-peak degradation of PT-MWs with or without confining plates. Using the validated procedure, a parametric study was performed to investigate the effect of different parameters on the compression zone length. Multi-variate regression analysis was then performed to develop an equation to evaluate the compression zone length. In the proposed equation, the compression zone length is expressed as a function of the wall length and axial stress ratio. This expression was incorporated into the flexural analysis of PT-MWs. The strength of 14 tested PT-MWs and two sets of FEMs of walls were compared with the values calculated using the proposed approach and the MSJC (2013) approach. According to the results, disregarding the elongation of the PT bars in the unbonded PT-MWs in the MSJC (2013) results in a highly conservative strength prediction. Comparatively, the proposed expression significantly improved the strength prediction. Based on the results of this study, the following expression is recommended to evaluate the flexural strength of PT-MWs, which is a simplified version of the equation proposed by Hassanli et al. (2014b):

$$M = \sum f_{ps\,i} A_{ps\,i} \left(d_i - \frac{\beta c}{2} \right) + N \left(\frac{L_w}{2} - \frac{\beta c}{2} \right) \tag{6.52}$$

where,

$$f_{psi} = f_{se\,i} + (0.00055 L_w + 17.375 \frac{f_m}{f_m'}) \frac{E_{ps}}{L_p} \left(\frac{d_i f_m'}{2 L_w f_m} - 1 \right) \tag{6.53}$$

References

Bean Popehn JR, Schultz AE, Drake CR (2007) Behavior of slender, post-tensioned masonry walls under transverse loading. J Struct Eng 133(11):1541–1550

Bohl A, Adebar P (2011) Plastic hinge lengths in high-rise concrete shear walls. ACI Struct J 108 (2):148–157

British Standards Institution (2005) Code of practice for the use of masonry. 2: structural use of reinforced and pre-stressed masonry. BS 5628-2, London

Dawood HM, ElGawady MA, Hewes J (2011) Behavior of segmental precast post-tensioned bridge piers under lateral loads. J Bridge Eng 17(5):735–746

Dawood HM, ElGawady MA (2012) Performance-based seismic design of unbonded precast post-tensioned concrete filled GFRP tube piers. Compos Part B

ElGawady MA, Sha'lan A (2011) Seismic behavior of self-centering precast segmental bridge bents. J Bridge Eng, ASCE 16(3):328–339

ElGawady MA, Booker AJ, Dawood H (2010) Seismic behavior of post-tensioned concrete-filled fiber tubes. J Compos Constr, ASCE 14(5):616–628

Harajli M (2006) On the stress in unbonded tendons at ultimate: critical assessment and proposed changes. ACI Struct J 103(6):1–10

Hassanli R, Elgawady MA, Mills JE (2014a). Strength and seismic performance factors of post-tensioned masonry walls. J Struct Eng (in press)

Hassanli R, Elgawady MA, Mills JE (2014b) Flexural strength prediction of post-tensioned masonry walls. J Struct Eng (under review)

Henry RS, Brooke NJ, Sritharan S, Ingham JM (2012) Defining concrete compressive strain in unbonded post-tensioned walls. ACI Struct J 109(1):101–111

Laursen PPT (2002) Seismic analysis and design of post-tensioned concrete masonry walls. Ph.D. dissertation, Department of Civil and Environmental Engineering, University of Auckland, Auckland, New Zealand

Masonry Standards Joint Committee (MSJC) (2008) Building code requirements for masonry structures, ACI 530/ASCE 5, TMS 402. American Concrete Institute, Detroit

Masonry Standards Joint Committee (MSJC) (2013) Building code requirements for masonry structures, ACI 530/ASCE 5, TMS 402. American Concrete Institute, Detroit

Mortezaei A, Ronagh HR (2012) Plastic hinge length of reinforced concrete columns subjected to both far-fault and near-fault ground motions having forward directivity. Struct Des Tall Spec Build 19(6):903–926

Paulay T, Priestley MJN (1992) Seismic design of reinforced concrete and masonry buildings. Wiley, Inc., 767

Paulay T, Priestly MJN (2009) Seismic design of reinforced concrete and masonry buildings. Wiley, Inc.

Priestley MJN, Elder D (1983) Stress-strain curves for unconfined and confined concrete masonry. ACI J Proc (ACI) 80(3):192–201

Priestley MJN, Calvi G, Kowalsky M (2007) Displacement-based seismic design of strutures. IUSS Press, Pavia, Italy

Rosenboom OA, Kowalsky MJ (2004) Reversed in-plane cyclic behavior of post-tensioned clay brick masonry walls. J Struct Eng 130(5):787–798

Rosenboom OA (2002) Post-tensioned clay brick masonry walls for modular housing in seismic regions. M.S. thesis, North Carolina State University, Raleigh, NC, USA

Ryu D, Wijeyewickrema A, ElGawady M, Madurapperuma M (2014) Effects of tendon spacing on in-plane behavior of post-tensioned masonry walls. J Struct Eng 140(4). CID:04013096

Shedid M, El-Dakhakhni W (2014) Plastic hinge model and displacement-based seismic design parameter quantifications for reinforced concrete block structural walls. J Struct Eng 140 (4):04013090

Shedid M, El-Dakhakhni W, Drysdale R (2010) Seismic performance parameters for reinforced concrete-block shear wall construction. J Perform Constr Facil 24(1):4–18

Wight GD, Ingham JM (2008) Tendon stress in unbonded post-tensioned masonry walls at nominal in-plane strength. J Struct Eng 134(6):938–946

Wight GD (2006) Seismic performance of a post-tensioned concrete masonry wall system. Ph.D. dissertation, Department of Civil and Environmental Engineering, University of Auckland, Auckland, New Zealand

Chapter 7
Experimental Investigation of Unbonded Post-tensioned Masonry Walls

This chapter reports on an experimental program conducted as a part of this thesis that investigated the behavior of PT-MWs. The accuracy of the MSJC (2013) in ignoring the elongation of PT bars is investigated using the two design equations proposed in Chaps. 5 and 6 to predict the flexural strength of the tested walls. The accuracy of the analytical approach developed in Chap. 6 is verified against the presented experimental results.

7.1 Introduction

Unbonded post-tensioning induces self-centering behavior in structural systems. In recent years, this technology has been incorporated in masonry walls to provide a rocking behavior (Laursen 2002; Rosenboom 2002; Wight 2006; Ryu et al. 2014; Hassanli et al. 2014a). Self-centering behavior helps to eliminate or reduce the residual drift and limits the structural damage. Hence, the structural system maintains its integrity following a ground excitation event and the damage zone is limited to the toe of the walls, which can be repaired with minimal cost.

The majority of past research on unbonded post-tensioned masonry walls (PT-MWs), focused on out-of-plane behavior. An extensive literature review of out-of-plane behavior of PT-MWs was carried out by Lissel et al. (2000) and Schultz and Scolforo (1991). However, experimental studies on the in-plane response of unbonded PT-MWs are limited. Page and Huizer (1988) conducted one of the early studies on PT-MWs and investigated the effect of horizontally and vertically pre-stressing clay masonry walls under monotonic loading. The first in-plane cyclic testing of PT-MWs was carried out by Rosenboom and Kowalsky

With permission from ASCE: Hassanli R., ElGawady M.A. and Mills J.E., Experimental investigation of cyclic in-plane behavior of unbonded post-tensioned masonry walls', Journal of Structural Engineering, 142(5), 2016.

(2004), who studied the effect of bonded and unbonded PT bars, as well as the effect of supplemental mild steel and confinement plates on the behavior of PT-MWs. Laursen and Ingham (2001, 2004a, b) performed extensive research on the behavior of post-tensioned concrete masonry walls under in-plane cyclic loading. The effects of aspect ratio, pre-stressing level, grout infill, location of PT bars and confinement plates were investigated through experimental testing of eight walls in phase I, five walls in phase II and two walls in phase III. They concluded that for unbonded confined and unconfined PT-MWs, the wall can reliably reach a drift of 1.0% and 1–1.5% without experiencing strength degradation. Moreover they found that the level of energy dissipation in unbonded PT-MWs is comparatively low and that the self-centering behavior was retained even after tendon yielding.

Ewing and Kowalsky (2004) carried out experimental tests on perforated PT-MWs and concluded that it is possible to design perforated unbonded post-tensioned clay brick masonry walls to maintain all of the benefits of solid PT-MWs, provided properly designed cold joints are included that divide a single unbonded perforated PT-MW into multiple piers. Shaking table tests of solid and perforated PT-MW specimens and simple square structures were also carried out by Wight (2006). This study verified the ability of such walls to return to their original vertical alignment and withstand large numbers of excitations with minimal damage. The damage of the simple structure was consistent with that obtained for perforated walls, being bond beam cracking and cracking below the openings (Wight 2006).

According to MSJC (2013) the minimum reinforcement required in the wall construction for it to be considered as a reinforced masonry wall is as follows: Vertical reinforcement of at least 0.129 m^2 in cross-sectional area needs to be provided at corners, within 203 mm of the ends of walls, and at a maximum spacing of 3048 mm on the wall centerline. Horizontal reinforcement must consist of at least two longitudinal wires of W1.7 (MW11) joint reinforcement spaced not more than 406 mm on center, or at least 0.129 m^2 in cross-sectional area of bond beam reinforcement spaced not more than 3048 mm center to center. For walls with openings or movement joints additional criteria are required for reinforcement by MJSC (2013). There are limited studies on whether or not a wall with unbonded PT bars should be considered as a reinforced masonry wall. Moreover, no studies have been done to investigate the self-centering and energy dissipation response considering the effect of arrangements of PT bars and the level of post-tensioning. A review of the literature indicates that the majority of tests on PT-MWs have been designed so that the PT bar yields.

The design approach for masonry provided by MSJC (2013) is completely different from the approach enforced by ACI 318-08 (2008) for concrete members. To ensure ductility in an unbonded masonry wall based on MSJC (2013) the PT bars need to be yielded. However, according to ACI, to design an unbonded concrete wall, the PT bars must remain elastic, to ensure the self-correcting response of the wall. Comparing the two approaches, it is therefore important to investigate whether or not the unbonded post-tensioned bar should be yielded. The experimental program was designed to answer this question, and to investigate the effect of PT bar spacing and level of initial axial stress ratio, f_m/f'_m, on the behavior of PT-MWs. Emphasis was made on: (1) The overall behavior of fully grouted PT-MWs on the basis of the experimentally recorded force-displacement hysteretic relationships. (2) The influence of the PT bar spacing and the level of initial stress on the wall demands. (3) Providing experimental evidence to support the proposed design methodology to predict the flexural strength and force-displacement response of PT-MWs (see Chaps. 5 and 6). (4) Developing design criteria for unbonded PT-MWs to ensure ductility and a self-centering response.

The following sections provide details of the materials, wall construction, test-set-up, test results and strength and force displacement predictions of the wall specimens.

7.2 Construction Details

7.2.1 Wall Specifications

A total of four unbonded PT-MWs were manufactured and tested under incrementally increasing displacement cyclic load applied laterally to the top of the walls. Fully grouted wall was considered rather than partially or ungrouted walls. Although, in terms of labor and material costs, construction time and added magnitude of seismic mass, ungrouted and partially grouted walls are more favorable than fully grouted walls, according to the previous test results, they exhibit a non-ductile response, characterized mainly by diagonal shear failure (Laursen 2002; Rosenboom and Kowalsky 2004).

The test walls had a thickness/height/length of 190 mm/2300 mm/1400 mm, corresponding to an aspect ratio (height/length) of 1.64. The 2300 mm height of the walls was measured from the mid-height of the loading beam to the top of the footing surface. Figure 7.1 illustrates the geometry of the specimens.

All walls were constructed using concrete masonry blocks with a nominal thickness, b_w, of 190 mm, height of 190 mm and length of 390 mm. The walls were 3.5 concrete masonry blocks long and 10 courses high, as shown in Figs. 7.1 and 7.2. The length of a typical wall included seven vertical cores which enabled different symmetrical configurations of PT bars. The walls were vertically post-tensioned using PT bars passing through vertical ducts located inside the walls. In the case of wall W1, two PT bars were placed in the end cells. Wall W2 was post-tensioned using three PT bars with two bars placed in the end cells and one in the middle cell, providing a horizontal spacing of 600 mm between the bars. Both walls W3 and W4 consisted of four PT bars located in every other vertical core, providing a spacing of 400 mm between the bars (Fig. 7.1). Wall W4 contained horizontal reinforcement and was subjected to a higher level of initial pre-stressing load compared with wall W3. The test matrix is presented in Table 7.1.

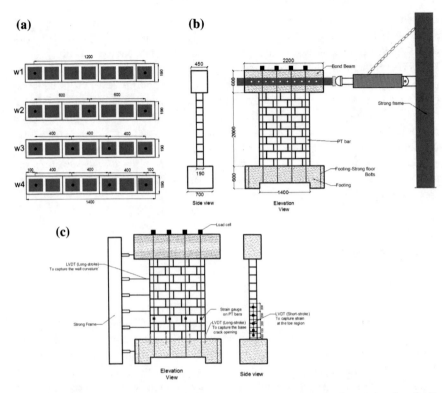

Fig. 7.1 Testing detail **a** cross sections, **b** details of the test setup and **c** instrumentation detail

Fig. 7.2 Test setup

Table 7.1 Test matrix and specimen details

Specimen	Thickness (m)	Length (m)	Height (m)	No. of PT bars	PT bar spacing (m)	PT bar initial force (kN)	f_i^* (MPa)	PT bar total area (mm²)	f_i/f_{py}	Pre-compression stress on the wall (MPa)	Shear reinforcement
W1	0.19	1.4	2.3	2	1.2	180	573	628	0.63	1.35	No
W2	0.19	1.4	2.3	3	0.6	120	382	942	0.42	1.35	No
W3	0.19	1.4	2.3	4	0.4	90	287	1256	0.32	1.35	No
W4	0.19	1.4	2.3	4	0.4	180	573	1256	0.63	2.71	2N20

f_i: initial stress in the PT bar after immediate stress losses

All the walls were post-tensioned with 20 mm diameter Dywidag PT bars placed in 32 mm diameter PVC ducts. Walls W1, W2, and W3 were post-tensioned with a total initial force of 360 kN shared between the PT bars, corresponding to an axial stress on the masonry of 1.35 MPa, while wall W4 was post-tensioned with a total initial force of 720 kN shared between the four PT bars, corresponding to an axial stress on the masonry of 2.7 MPa. It worth noting that walls W1, W2 and W3 were designed to provide a flexural response. As the shear capacity of the walls without shear reinforcement was estimated to be greater than the flexural capacity, no horizontal shear reinforcement was used in these walls. However, Wall W4 included horizontal reinforcement, which was provided in the fourth and seventh courses. Each level consisted of a single N20 rebar (area = 314 mm^2, yield strength = 500 MPa) being approximately 1300 mm long with a 180° hook around the outermost vertical PT bar duct. The webs of the blocks in courses that included horizontal reinforcement were cut and knocked-out to a depth of 110 mm to accommodate the horizontal reinforcement.

7.2.2   Wall Construction

A professional mason constructed the test specimens using concrete blocks (CMU) and Portland cement lime type S mortar formulated using 1: 1: 6 (cement: lime: sand) by volume. The walls were constructed in standard running bond pattern using face shell mortar bedding with a mortar thickness of 10 mm. Each wall panel was constructed on a 0.7 m thick by 0.6 m height by 2.2 m length precast concrete footing. The footing was recessed to be able to access the bottom anchorage to assemble the post-tensioning nuts and plates.

A reinforced concrete loading beam having dimensions of 0.45 m thick by 0.6 m height by 2.2 m length was placed on top of each wall using a layer of high-strength mortar with a thickness of 10 mm. Both the precast reinforced concrete footing and loading beam were manufactured by a specialist precast concrete company to the researcher's design, using concrete having a compressive strength of 45 MPa.

Construction of the walls proceeded as follows: The PVC tubes were placed and fixed in position in the footing and the first four courses were constructed. For wall W4, the horizontal shear reinforcement was placed. The walls were then fully grouted two days after construction using a high slump grout with a maximum aggregate size of 10 mm. The same processes were carried out for courses 5–7 and 8–10. To fill the gap between the wall and loading beam and to provide a uniform bearing surface for stress transfer, a layer of high-strength mortar with a thickness of 10 mm was placed between the wall and the top bond beam. The wall-footing system was moved to the testing rig after 210 days, the loading beam was placed on top of the wall, the PT bars were installed in the PVC ducts and then the PT bars were stressed.

7.3   Material Properties

Standard 200-mm hollow concrete blocks (190 mm depth × 190 mm width 390 mm length) were used to construct the test walls, which are similar to 8 inch blocks commonly used in the United States. A series of material tests were performed to determine the properties of blocks, mortar and concrete according to ASTM standards.

The average compressive strength of the CMUs, tested according to ASTM C140 (2013) was 19.5 MPa with a standard deviation (STD) of 0.9 and a coefficient of variation (COV) of 4.6%. The average compressive strength of a grout cylinder was 23.0 and 26.8 MPa at 28-days and the day of wall testing, respectively. The average compressive strength of grout cubes prepared according to ASTM C1019 (2011) was 28.7 MPa at the day of wall testing. The indirect (Brazilian) tensile strength (ASTM C496 2011) of grout cylinders was 2.23 MPa (COV = 4.1% and STD = 0.09).

A total of 15 two-block high by one-block long prisms were constructed using PCL type S mortar. These prisms were then tested according to ASTM C1314 (2012) to determine the masonry compressive strength. The average compressive strength of the grouted masonry prisms, f'_m, at the day of wall testing was 17.5 MPa with a STD of 0.3 and COV of 1.7%.

The nominal properties of the PT bars were tensile yield stress of f_{py} = 900 MPa, corresponding to a yielding load of 283 kN and ultimate tensile stress of f_{pu} = 1100 MPa, corresponding to an ultimate load of 346 kN.

7.4   Instrumentation

Displacement potentiometers were used to measure the wall horizontal displacements at various heights, vertical displacements, and horizontal movement of the footing (Fig. 7.1c). Electrical strain gauges were used to monitor the strain developed in the PT bars. Load cells were also employed near the live end anchorage point to record the forces in the PT bars.

The stress-strain behavior of the PT bars was obtained during the tests using load cells and strain gauges. A bi-linear idealization of the PT bar was conducted and resulted in average yield strength of 903 MPa, Young's modulus of 190.4 GPa and post-yield stiffness of 32 MPa.

7.5   Test Setup and Test Procedure

Wall testing was conducted using the setup shown in Figs. 7.1 and 7.2. The lateral cyclic load was applied using a 1000 kN hydraulic jack having a maximum stroke of ±75 mm. A guidance frame was employed to prevent out-of-plane movement of

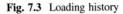

Fig. 7.3 Loading history

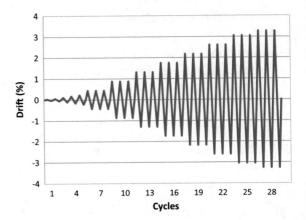

the loading beam and wall. The guidance frame (Fig. 7.2) was pinned to the loading beam at one end and fixed to the strong floor at the other end.

All walls were subjected to identical cyclic in-plane loading applied to the mid-height of the top loading beam, providing an effective wall height of 2.3 m corresponding to an effective aspect ratio of 1.64. The testing procedure was based on the ASTM E2126 (2011) standard test method B for a load test that includes the following two displacement patterns: (1) Five single fully reversed cycles at displacement amplitudes of 1.25, 2.5, 5, 7.5 and 10% of the ultimate displacement, (2) Three fully reversed cycles at displacements of 20, 40, 60, 80, 100, and 120% of the ultimate displacement. The ultimate displacements were estimated as 75 mm using finite element models (Hassanli et al. 2014b). All walls were taken through similar cyclic displacement regimes as shown in Fig. 7.3. Due to the accidental misplacement of wall W4 in the testing rig before starting the test, the wall was not located at the center of the testing rig, hence due to limitations of the stroke of the ram, a maximum displacement of 45.5 and 82.3 mm was applied to the wall top in push and pull directions (75 and 110% of ultimate displacement respectively).

7.6   Test Results

7.6.1   Damage Pattern and Failure Mode

The test specimens displayed a rocking response characterized by opening of a single large crack at the wall-foundation interface. No other flexural cracks were observed during testing (Fig. 7.4a, c, e, g, h). Shear cracks were observed in walls W1, W2 and W3 toward the end of the tests and after the walls experienced large displacements.

At a drift ratio of 0.24–0.30%, a small crack was observed at the wall-footing interface. This crack became more significant and reached about half the length of

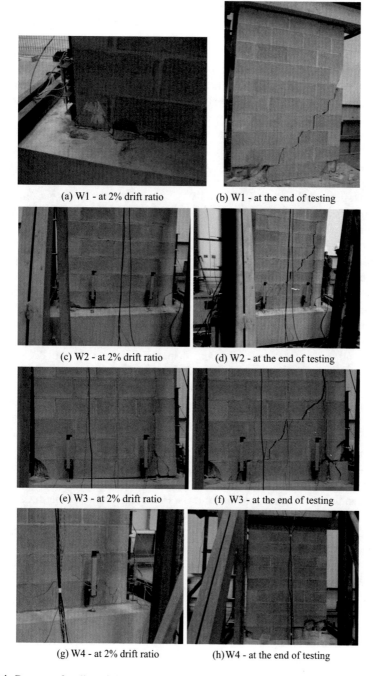

(a) W1 - at 2% drift ratio   (b) W1 - at the end of testing

(c) W2 - at 2% drift ratio   (d) W2 - at the end of testing

(e) W3 - at 2% drift ratio   (f) W3 - at the end of testing

(g) W4 - at 2% drift ratio   (h) W4 - at the end of testing

Fig. 7.4 Damage of walls at 2% drift ratio and end of testing

the wall at a drift ratio of 0.58–0.68%. Minor vertical cracks were observed in the bottom-most concrete blocks at the wall's toe at a drift ratio of about 0.93–1.04%. At a drift ratio of 1.37–1.54%, these vertical cracks extended along the bottom-most block height. At a drift ratio of 1.82–1.98%, spalling of masonry face shells was observed. Until this drift ratio, the behavior of all walls was approximately the same. However, some hairline vertical cracks were formed at 200 mm distance from the edges of walls W2 and W3 at a drift ratio of 1.9%. The cracks started from the wall-footing interface joint and extended to the second course.

Figure 7.4a, c, e, g show the damage of the walls at a drift ratio of 2%, and Fig. 7.4b, d, f, h show the wall damage at the end of testing. Prior to a drift ratio of 2% all walls presented a flexural response. By a drift ratio of 2.2–2.4%, the cracking and spalling at the compression toe region became more pronounced in all walls and the vertical cracks in walls W2 and W3 extended to the fourth masonry course (Fig. 7.4c, e). At a drift ratio of 2.6%, walls W2 and W3 suddenly failed due to a major diagonal shear crack that extended from the mid-height of the wall to the compression zone region (Fig. 7.4d, f). The same mechanism of a diagonal crack was observed for wall W1 by a drift ratio of 3.0% (Fig. 7.4b). The sudden shear failure can be attributed to the incapability of the walls to undergo large deformations without any shear reinforcement. The presence of shear reinforcement in wall W4 prevented the occurrence of the vertical and diagonal cracks which formed in the other walls. Hence, wall W4 exhibited a flexural response throughout the whole testing regime.

As mentioned, a maximum unequal displacement of 45.5 and 82.3 mm was applied to the wall W4 in push and pull directions, respectively. This caused the extent of the damage to be confined to the first course and the first two courses in pull and push directions, respectively (Fig. 7.4h). In the push direction, after the degradation and failure of the masonry at the toe region, the compression zone migrated towards the center of the wall, characterized by an extended damage zone along the length of the wall.

The flexural failures of the walls were signified by spalling of masonry at the toe region (Fig. 7.4a, c, e, g, h). As shown in the figures, the extent of damage was confined to the lowest two masonry courses. However, gradual strength degradation was not observed except in wall W4. It is likely that the walls W1, W2 and W3 would display a more gradual strength degradation should shear reinforcement be provided similar to wall W4. Therefore, it seems that minimal shear reinforcement is required to maintain the integrity of the wall under high levels of drift (more than 2% in this study).

7.6.2 Force-Displacement Response

The force-displacement responses of the walls based on the experimental data are presented in Fig. 7.5. The "analysis" line in these graphs was obtained using the simplified approach developed in Chap. 6 and will be discussed in more detail later

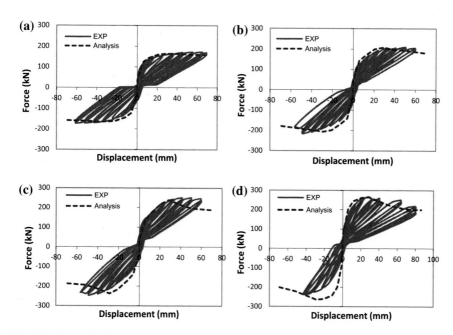

Fig. 7.5 Force-displacement experimental results and theoretical prediction of walls a W1, b W2, c W3 and d W4

in the chapter. Very stable hysteretic loops were observed for all walls. Walls W1, W2, W3 and W4 withstood applied horizontal loads of 172.4, 209.4, 250.9 and 266.6 kN in the push direction and 172.9, 217.7, 247.5 and 240.6 kN in the pull direction, respectively. As shown in Fig. 7.5d in wall W4 a gradual strength degradation was observed in the push direction. As presented in Fig. 7.5, walls W1, W2 and W3 developed a large displacement capacity of more than 46 mm corresponding to a drift ratio of more than 2% without showing significant strength degradation. The strength of Wall W4 peaked at a displacement of 28.7 and 39.4 mm, corresponding to a drift ratio of 1.25 and 1.72% in push and pull directions, respectively. As shown in Fig. 7.5d wall W4 was the only wall that experienced gradual strength degradation and reached 80% of ultimate strength at a displacement of 80.6 mm corresponding to a drift ratio of 3.5% (push direction).

The comparison of the force-displacement envelopes (backbone of the force-displacement hysteretic response) of the four walls is provided in Fig. 7.6. The addition of one and two PT bars in walls W2 and W4 with respect to wall W1 caused an average in-plane strength increase of 24 and 44%, respectively.

Comparing the force-displacement response of the walls W3 and W4 in Fig. 7.6, it is seen that on average wall W4 developed only 2% higher strength than W3, even though the axial PT stress in wall W4 was double that in wall W3. This behavior differs to the test results reported by Laursen and Ingham (2004b). Comparing the experimental results of two large-scale PT-MWs, walls S3-1 and

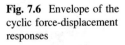

Fig. 7.6 Envelope of the cyclic force-displacement responses

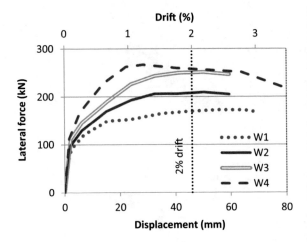

S3-2, Laursen and Ingham (2004b) demonstrated that by increasing the axial stress ratio by 29% the strength increased by 21%. In other words, the effect of axial stress on the in-plane strength of their walls was quite significant compared to the results presented in this study. The difference in these outcomes can be attributed to the yielding of the PT steel. While in all test walls reported here (walls W1 to W4), the extreme PT bars yielded, in walls S3-1 and S3-2, the PT steel remained elastic during testing. Therefore, in walls S3-1 and S3-2, the in-plane strength was governed by crushing of concrete; whereas the yielding of the PT bars of walls W3 and W4 governed the strength, resulting in approximately the same in-plane strengths exhibited by walls W3 and W4, regardless of the level of axial stress ratio.

According to the International Building Code (IBC 2006), to control damage to the life safety level, the inter-story drift under the design level earthquake should not exceed 2%. As can be seen in Fig. 7.6, the drift capacity of all walls was beyond 2% without considerable strength degradation. According to Fig. 7.6, while wall W1 underwent 3% drift before it failed in shear, shear failure of the walls W2 and W3 occurred at a lower drift of 2.6%. This can be attributed to the existence of intermediate PT bars in walls W2 and W3 compared to Wall W1. Developing extra stress in the intermediate bars during testing applied additional axial stress to the walls W2 and W3 compared with wall W1, and resulted in a more brittle failure due to the increased axial stress at the end of testing.

7.6.3   Force in the PT Bars

The total PT bar force of each wall obtained from the load cells is plotted against the in-plane displacement in Fig. 7.7. As mentioned, the initial total PT forces were 360 kN in walls W1, W2 and W3 and 714 kN in wall W4. During the rocking mechanism and due to the opening of the interface joint crack at the wall base, the

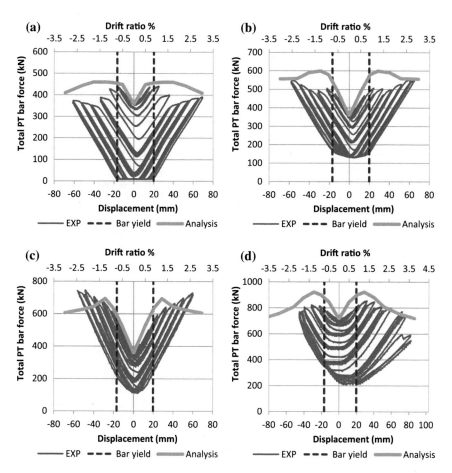

Fig. 7.7 Total post-tensioning force versus top displacement of walls **a** W1, **b** W2, **c** W3 and **d** W4

PT bars elongated and the force in the PT bars significantly increased. The forces reached maximum values of 439, 554, 718 and 852 kN representing increases of 22, 54, 100, and 19% for walls W1, W2, W3, and W4, respectively. As shown in Fig. 7.7, the PT bars in walls W1, W2, W3, and W4 yielded at drifts of 0.86, 1.12, 1.52 and 1.27% in push and drifts of 0.73, 1.08, 1.15 and 1.29% in pull directions, respectively. The peak force in the extreme tension PT bar of walls W1 to W4 reached a maximum of 379, 367, 364 and 340 kN, respectively, all of which were beyond the yielding point.

As shown in Fig. 7.7, during the unloading portion of the tests the PT force in the bars reduced. The losses in PT bar force occur through the following three mechanisms: (1) Yielding of PT bars due to stressing the PT bars beyond the elastic limits; (2) Degradation of masonry stiffness and toe crushing resulting in shortening of the wall; (3) Anchorage losses caused by deformation and movement of the PT plates and nuts. The PT force of walls W1, W2, W3 and W4 decreased to zero, 130, 114 and 207 kN corresponding to 100, 64, 68 and 42% reduction in the PT forces, at the end of testing, respectively.

Figure 7.8a shows the envelope of the total PT force, normalized by the initial PT force, versus displacement. The envelope of the PT force is obtained using the first loop at each drift ratio level. As shown in the figure, in general, the PT force increases with applied displacement. While in wall W1 and W4 the ratio reached a maximum of 1.20 and 1.18 respectively, it increased to a value of 1.53 and 2.01 in walls W2 and W3, respectively. Walls W1 and W4 exhibited a reduction in total PT force after reaching a drift of 1.04 and 1.35% in the push direction, and 1.11 and 1.38%, in the pull direction, respectively. This behavior can be attributed to the level of f_i/f_{py} in the test walls. According to Table 7.1, the initial to yield stress ratio

Fig. 7.8 **a** Normalized enveloped PT force, and **b** residual post-tensioning force ratio

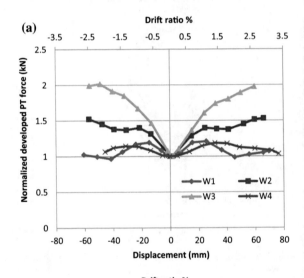

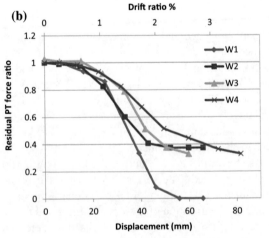

in the PT bar, f_i/f_{py}, was 0.63, 0.42, 0.32 and 0.63 in walls W1, W2, W3 and W4, respectively. The higher ratio of f_i/f_{py} in walls W1 and W4 compared to walls W2 and W3, resulted in yielding of the PT bars and relevant losses at lower drifts. It can therefore be concluded that as the f_i/f_{py} ratio increases, the total PT force developed in the PT bars decreases.

7.6.4 Residual PT Force Ratio

Self-centering behavior of an unbonded PT-MW can be achieved if the loads in the PT bars can return the wall back to its original alignment. Hence, it is important to estimate the amount of residual PT force retained in the PT bars after each cycle of test. To better understand the losses that occur in PT-MWs during cyclic loading, Fig. 7.8b presents the residual PT force recorded at the end of the first unloading cycle at each drift level for the test specimens. The residual PT force ratio was defined as the ratio of the total residual PT force in the third cycle of loading to the total initial PT force. As can be seen in the figure, there are no significant PT losses prior to a drift ratio of 0.65%. Within this drift ratio limit, none of the PT bars had yielded and the small losses can be attributed to minor tensile cracks in the masonry, as well as small relative movements of the PT plates and nuts at the anchorage. Loss of PT force occurred rapidly at a drift ratio of about 1–2%. The PT force degradation in this drift range can be attributed to yielding of the bars and permanent elongation of the extreme tensile PT bars.

Wall W1 was the only wall in which the residual PT ratio reached zero. Interestingly, as shown in Fig. 7.5a, this wall was the only wall which experienced a zero-stiffness region in the force-displacement response. It seems that the complete loss of pre-stressing force in wall W1 led to the presence of the zero-stiffness response and potential sliding. Walls W2, W3 and W4 retained 29–36% of their initial total PT force at the end of wall testing and no zero–stiffness region was recorded for these walls. To illustrate the development of stress in the tested specimens, Fig. 7.9 presents the force developed in each PT bar of wall W3. According to the figure, the PT force in the extreme tension PT bar reached the yield stress of 283 kN, however, the stress in the intermediate PT bars remained below the yielding point. As shown, during unloading, there was a gradual degradation in the PT force. However, when the specimen was loaded to a drift ratio of 2.18 and 2.15% in push and pull directions, respectively; unloading during this cycle resulted in zero PT forces. Similar behavior was observed for the extreme bars in all walls with zero PT force when loading exceeded the drift ratio of 2.42, 2.3 and 2.57% in push, and 1.87, 1.81 and 1.96% in pull direction, for walls W1, W2 and W4 respectively. As shown in Fig. 7.9, for wall W3, the central PT bars retained 55–65% of their initial PT force at the end of testing. The same mechanism occurred in walls W2 and W4, which also included intermediate PT bars along the wall length. In wall W2 and W4, 98 and 60% of the PT force was retained in the

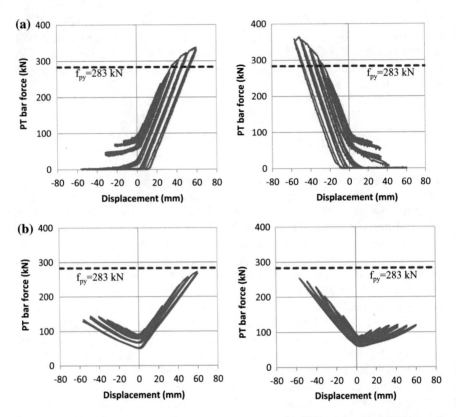

Fig. 7.9 Force developed in Wall W3 in the **a** extreme tension PT bars, **b** central PT bars

central PT bars at the end of testing. The residual PT force in the central PT bars in walls W2, W3 and W4 provided the self-centering response and prevented a zero-stiffness region in the force-displacement response.

7.6.5 Residual Drift Ratio

The lateral drift at the point of zero lateral force at the end of the first cycle of each imposed lateral displacement level was considered as the residual drift, and is presented versus displacement for each wall in Fig. 7.10. Before a drift ratio of 0.6% the residual drift was lower than 0.13% for all walls, and was approximately the same for all walls. However, after this point, while the residual drift was stable in walls W2, W3 and W4 and remained lower than 0.15%, wall W1 exhibited an increased residual drift. According to Fig. 7.10, at a lateral imposed drift of 2 and 3%, the residual drift of wall W1 reached 0.29 and 0.83%, respectively. This behavior can be attributed to the yielding of all PT bars which resulted in zero stiffness in the force-displacement response.

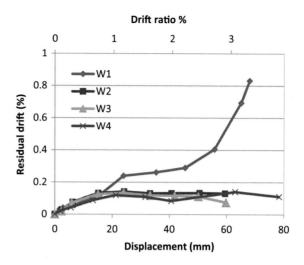

Fig. 7.10 Residual drift

7.6.6   Strain in Masonry

The change in masonry vertical displacement along the bottom-most 250 mm at the wall's toe and heel was recorded during testing using LVDTs. These changes in displacements were converted to average vertical strains in the masonry and presented in Fig. 7.11 for pull and push directions. Strain measurements presented in the figure reveal that the strain in the concrete masonry in unbonded PT-MWs could reach high values of 0.03–0.04 before strength degradation occurs. Even though none of the walls, except W4, experienced strength degradation due to the degradation of the masonry strength at the toe region, high levels of strain were recorded during the experiment. When the maximum strength of the walls is governed by the

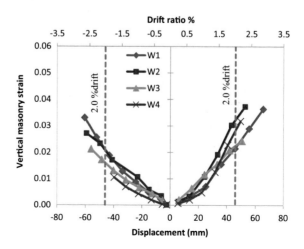

Fig. 7.11 Maximum compressive strain in masonry recorded in different displacement cycles

yielding of PT bars, as occurred in the test walls, degradation in masonry strength in the toe region causes a reduction in the in-plane strength. As shown in Fig. 7.11, at a drift ratio of 2%, which is commonly considered by codes as the ultimate drift limit state for the design of unbonded PT walls, the masonry strain of the walls reached a value of 0.016–0.032. This range of strain is considerably higher than the value of 0.003 commonly used for bonded PT walls. This is in line with the conclusion reported by other researchers, that the strain in unbonded walls reached higher values compared with that in bonded walls (Henry et al. 2012; Laursen and Ingham 2004b).

The vertical strain profile in the masonry was obtained through a series of LVDTs attached to the edge of the wall (Fig. 7.1c). Figure 7.12 presents an example of the strain which developed along the height of wall W1. A steep strain increase is detectable at the compression region. This was similar in all tested walls. The summation of the vertical strains at the tension and compression edge of the wall, measured by externally mounted LVDTs, was divided by the wall length to get the average curvature of each wall (Fig. 7.13). According to Figs. 7.12 and 7.13 the strain and curvature were highest in the lower part of the walls, mainly below 400 mm wall height, corresponding to the first two courses. This is compatible with the extent of flexural damage observed in the experiment.

7.6.7   Compression Zone Length and Wall Rotation

The linear potentiometers attached along the wall-footing interface were used to capture the base crack opening profile (Fig. 7.1c), plotted in Fig. 7.14. At a drift ratio of 2%, the wall gap opening reached approximately 20 mm.

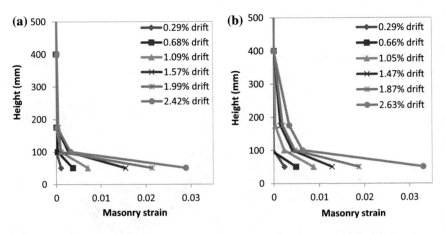

Fig. 7.12 Strain profile along the height of wall W1 in **a** push direction, and **b** pull direction

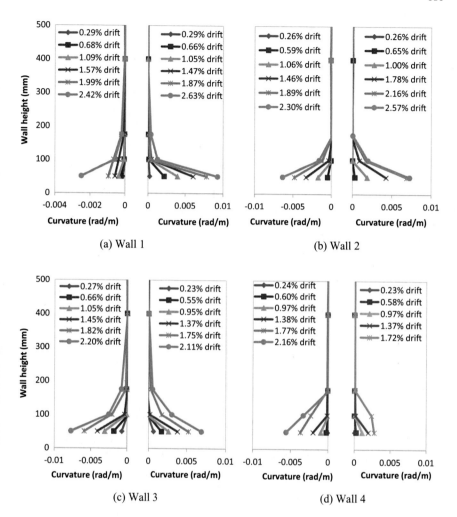

Fig. 7.13 Average curvature profile along the wall height

These graphs were used to determine the compression zone length, c, and the wall rotation at the base, presented in Figs. 7.15 and 7.16, respectively. The compression zone length remained approximately the same for all displacements, at approximately 30% of the wall length. Using an analytical approach and employing multi-variate regression analysis, the following equation was proposed by Hassanli et al. (2014c) to estimate the compression zone length in unbonded PT-MWs, (as described in Chap. 6),

$$c = 2l_w \frac{f_m}{f'_m} \tag{7.1}$$

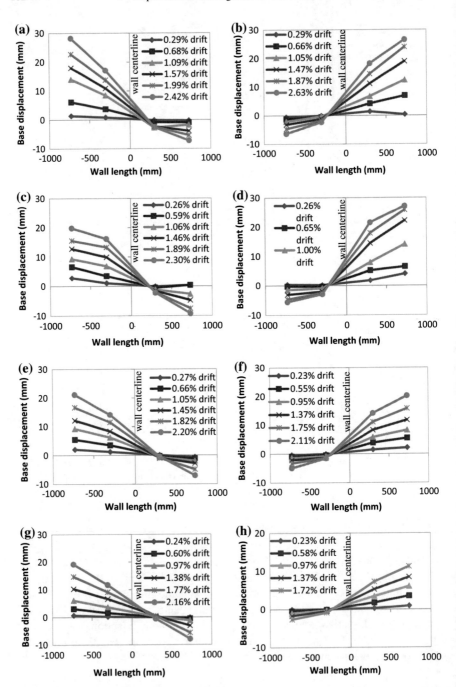

Fig. 7.14 Base crack profile of wall **a** W1-push, **b** W1-pull, **c** W2-push, **d** W2-pull, **e** W3-push, **f** W3-pull and **g** W4-push, and **h** W4-pull

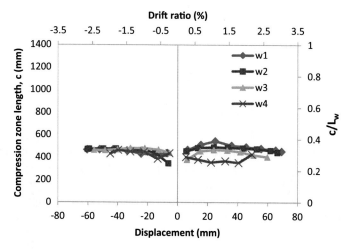

Fig. 7.15 Compression zone length, c

Fig. 7.16 Drift of the wall
versus wall rotation at the
base of walls

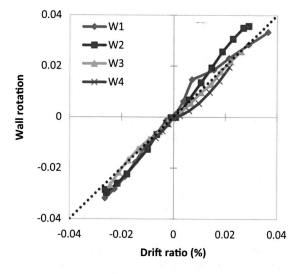

The compression zone length estimated using Eq. 7.1 was determined to be equal to 16–32% of the test wall length. Comparing these values with the experimental results presented in Fig. 7.15, shows that Eq. 7.1 tends to underestimate the compression zone length.

Figure 7.16 presents the wall rotation versus the drift ratio. The wall rotation was calculated assuming rigid body rotation of the wall around the middle of the compression zone and using the measured vertical displacements at the wall base i.e. the same data that was used to calculate the crack profiles at the base (Fig. 7.14). Drift ratio was calculated as the top wall displacement divided by the height of the wall. As shown in Fig. 7.16, the drift ratio and the wall rotation values were approximately the same, indicating that unbonded PT-MW systems respond mainly in rigid body rocking and hence flexural and shear deformation can be ignored.

7.6.8 Displacement Ductility

To evaluate the seismic parameters including response modification factor, R, displacement amplification factor, c_d, ductility, displacement ductility, μ and effective yield stiffness, K_e, the capacity curves were idealized by bilinear curves. The bilinear approximation of the capacity curve was performed using the process provided by FEMA 356 (2000). According to FEMA 356 (2000) the idealized relationship is bilinear with an initial slope and a post yield slope evaluated by balancing the area above and under the capacity curve. In this method, the initial effective slope at the base shear force is equal to 60% of the nominal yield strength and an iterative procedure must be used to balance the area. The bilinear approximation curves of the test walls are provided in Appendix C and the calculated seismic parameters are presented in Table 7.2.

As presented in the table, the yield displacements of the walls were 3.1, 3.5, 3.4, and 2.5 mm, for walls W1, W2, W3 and W4, respectively. While wall W1 presented the highest ductility value of 18.1, it was equal to 14.6, 14.4 and 11.2 in walls W2, W3 and W4. According to the table, all walls presented a comparatively high range of ductility, which can be attributed to the bi-linear elastic response of the unbonded PT-MWs. The higher level of axial stress in wall W4 compared to the other walls resulted in the lowest ductility value of the four walls.

7.6.9 Wall Stiffness

Stiffness properties of a wall can affect the fundamental period, displacement and distribution of lateral loads applied to a structure. For serviceability calculations, the initial stiffness estimation is of high importance. The initial stiffness is calculated as the slope of the load displacement curve for small drifts of 0.1% (before the

Table 7.2 Bilinear idealized curves parameters

Wall	Δy (mm)	Δu (mm)	V_y (kN)	V_u (kN)	K_e (kN/ mm)	αK_e (kN/ mm)	V_y/ V_u	μ
W1	3.1	55.3	134.4	172.4	44.0	0.7	0.8	18.1
W2	3.5	50.1	157.9	209.4	45.5	1.1	0.8	14.4
W3	3.4	50.2	178.6	250.9	52.0	1.5	0.7	14.6
W4	2.5	28.3	175.6	266.6	69.5	3.5	0.7	11.2

decompression point). For walls W1 to W4 this was determined to be equal to 60.2, 55.9, 64.7 and 86.0 kN/mm, respectively. The measured initial stiffness can be compared with the theoretical uncracked stiffness (k_g) considering shear and flexural deformation calculated using the following expression:

$$k_g = \frac{1}{\frac{h^3}{3E_m I} + \frac{h}{\alpha A_n G_m}} \tag{7.2}$$

where E_m and G_m are the modulus of elasticity and shear modulus, h is the wall height, I is the gross moment of inertia, A_n is the cross sectional shear area, α is the shape factor which accounts for the distribution of shear stresses across the section and is equal to 0.83 for rectangular sections. Equation 7.2 considers the flexural and shear deformation of the section. According to this equation, the initial stiffness theoretically depends only on the wall dimensions and elastic properties of masonry, which were the same for all tested walls. Hence, the theoretical stiffness for all walls was determined to be equal to 134.3 kN/mm. The measured initial stiffness was much lower than the theoretical uncracked stiffness, ranging from 42 to 64% of the theoretical value. Doubling the axial load in wall W4 compared to wall W3 caused an increase of 39% in the initial stiffness. It seems that to predict the initial stiffness of an unbonded PT member theoretically, the effect of axial load must also be considered.

The effective yield stiffness of the test walls, k_e, was determined using bi-linear idealized curves and was equal to 44.0, 45.5, 52.0 and 69.5 kN/mm, for walls W1, W2, W3 and W4, respectively, as presented in Table 7.2. To estimate the value of k_e of concrete members theoretically, Adebar et al. (2007) recommended an equation for the lower-bound. For masonry members Eq. 7.3 has been shown to provide an appropriate predictor of the yield stiffness (Banting and El-Dakhakhni 2014).

$$k_e = k_g \left(0.2 + 2.5 \frac{f_m}{f'_m} \right) \leq 0.7 \tag{7.3}$$

Using Eq. 7.3 the estimated k_e value was 53.9 for walls W1, W2, and W3; and 79.9 for wall W4. This corresponds to an overestimation of 23, 17, 4 and 15%, respectively. Using the test results presented here, the following modified equation was developed to predict the effective stiffness of unbonded PT-MWs.

$$k_e = k_g \left(0.15 + 2.5 \frac{f_m}{f'_m} \right) \leq 0.7 \tag{7.4}$$

The predicted k_e value according to Eq. 7.4 was 47.2 for walls W1, W2 and W3; and 73.2 for wall W4, all of which are within 10% of the experimental results.

7.6.10 Stiffness Degradation

To calculate the secant stiffness in each cycle of the test, the recorded force and displacements at the end of each half cycle were used. Although approximate, the secant stiffness was used to provide a qualitative comparison of stiffness degradation in the walls. Figure 7.17a presents the secant stiffness of the specimens versus displacement. As shown, walls W1 and W4 had the lowest and highest secant stiffness, respectively. Wall W4 presented a rapid reduction in secant stiffness, especially at small drift percentages. As shown in the figure, while doubling the axial load in wall W4 compared to other walls considerably increased the initial stiffness, at higher levels of displacements, its effect on the stiffness was reduced.

To reproduce the realistic behavior of a specimen in nonlinear modelling, an accurate estimation of the stiffness degradation is required. The stiffness degradation of bonded members can be attributed to the nonlinear behavior of material, however, in unbonded members the stiffness degradation behavior is highly influenced by the rocking mechanism. The normalized stiffness, represented as the secant stiffness divided by the yield stiffness, at different displacement ductility levels is presented in Fig. 7.17b. As shown in the figure, the test specimens demonstrated an approximately similar normalized stiffness degradation response. Using regression analysis of the test walls, the following expression was developed to predict the stiffness degradation of the unbonded PT-MWs,

$$\frac{k}{k_e} = 1.8D^{-0.8} \tag{7.5}$$

where D is the drift ratio of the wall. Figure 7.17b shows the comparison between the experimental data and the proposed equation. As shown in the figure, the proposed equation could effectively predict the stiffness degradation of the walls. It worth noting that Eq. 7.5 was developed using the test results of the four tested walls. More experimental results of walls with different configurations are required to verify its accuracy.

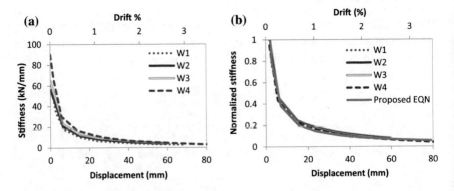

Fig. 7.17 Variation of **a** secant stiffness, and **b** normalized stiffness, versus top displacement

7.6.11 Equivalent Viscous Damping and Energy Dissipation

The equivalent viscous damping ratio, ξ_{eq}, is a parameter that defines the hysteretic damping behavior of the system. This parameter can be obtained by equating the elastic strain energy dissipated by a system with the energy dissipated from non-linear behavior (Jacobsen 1930), and can be calculated as follows:

$$\xi_{eq} = E_d / (4\pi E_s) \tag{7.6}$$

where E_s is the stored strain energy and E_d is the dissipated energy.

E_d was calculated as the area enclosed by a full cycle of response at each drift level (the first cycle was used in this study).

The calculated equivalent viscous damping ratio is plotted against the wall drift in Fig. 7.18a. As shown in the figure, the equivalent viscous damping in each loop was relatively small (about 5%). The reason can be attributed to the rigid body rocking and self-centering behavior of the system in which loading and unloading occurs approximately through the same path.

However, considering a single hysteresis loop is sensitive to the imposed load-displacement regime, the number of cycles prior to the current cycle and the displacement increment between consequent cycles. Therefore, researchers recommended considering an envelope of the previous load displacement hysteresis loops for a particular drift level (Hose and Seible 1999).

Figure 7.18b shows the equivalent viscous damping ratio (obtained from the envelope of the loops) versus displacement. The upward trend of ξ_{eq} reveals that in unbonded PT-MWs the range of values of damping can be considered in unbonded PT-MWs according to the design limit states and target displacement. For example, according to Fig. 7.18b at a drift ratio of 2%, the ξ_{eq} value is about 15–20%, which implies that unbonded PT-MWs can be expected to provide high levels of damping, and hence reduce the seismic deign loads, if the PT bars are designed to yield.

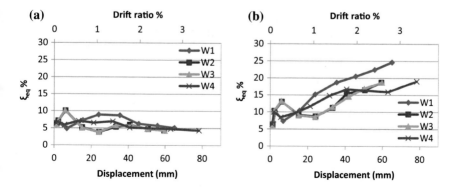

Fig. 7.18 Equivalent viscous damping **a** in each loop, **b** envelope of the previous loops

Fig. 7.19 Cumulative energy
dissipation

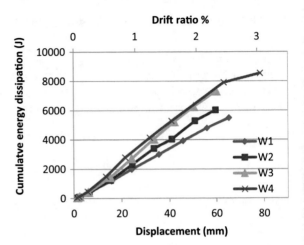

The accumulative energy dissipation was calculated by using the area inside the hysteretic loops. Figure 7.19 presents the cumulative energy dissipation for different displacements. As shown, Wall W1 dissipated the lowest amount of energy. Walls W3 and W4 dissipated the highest amount of energy, which was approximately the same at different displacements. The dissipated energy of these walls was approximately 45 and 20% higher than that of walls W1 and W2, respectively.

7.7 Response Prediction

In this section, the force-displacement response and the flexural strength prediction of the walls are provided and compared with the experimental results.

7.7.1 Force-Displacement Response Prediction

Hassanli et al. (2014c) presented a modification to the analytical approach originally proposed by Rosenboom and Kowalsky (2004) to predict the force-displacement response of unbonded PT-MWs. This approach was used here to predict the force-displacement responses of the tested specimens (the Analysis lines in Fig. 7.5). As shown in Fig. 7.5, the model can accurately predict the wall strengths, initial stiffness, and rotational capacities. The predicted strengths of the specimens using the analytical approach were 164, 207, 238 and 265 kN for walls W1, W2, W3, and W4, respectively, all of which fall within ±5% of the average strengths obtained in the test in pull and push directions.

One of the limitations of the proposed approach is that it is based on flexural response only and hence the shear failure that occurred toward the end of the tests can't be captured using the proposed approach. Figure 7.7 compares the total PT force developed in the PT bars obtained from the test and the analytical approach. As illustrated in the figure, the analytical approach slightly over predicted the force developed in the PT bars. This can be attributed to the losses in the PT forces that occurred during the test, due to the shortening of the wall and deformation of the anchorage, which were not considered in the analytical approach.

7.7.2 Strength Prediction of PT-MWs

MSJC (2013) has no procedure for estimating f_{ps} for unbonded PT-MWs. As mentioned before, according to MSJC (2013), instead of a more accurate determination, f_{ps} for members with unbonded pre-stressing bars can conservatively be taken as f_{se}. To investigate the accuracy of this approach, the flexural strength of the tested walls were calculated and were compared with that obtained from experimental results. The flexural strengths of the walls were also calculated using the recently developed simplified and iterative method of Hassanli et al. (2014c).

Masonry Standard Joint Committee (MSJC 2013)
Ignoring the elongation of PT bars, MSJC (2013) uses the following equation to predict the flexural strength of PT-MWs:

$$M_n = \left(f_{se}A_{ps} + f_yA_s + N\right)\left(d - \frac{a}{2}\right) \tag{7.7}$$

$$a = \frac{f_{se}A_{ps} + f_yA_s + N}{0.8f'_m b_w} \tag{7.8}$$

where a is the depth of the equivalent compression zone, A_s is the area of conventional flexural reinforcement, f_y is the yield strength, A_{ps} is the area of the PT bar, N is the gravity load including the self-weight of the wall, and d is the effective depth of the wall. The predicted in-plane strength of PT-MWs using this flexural expression is equal to the nominal moment capacity, M_n, divided by the effective height, h_n.

PT-MWs having no shear reinforcement can be considered as unreinforced masonry walls and hence the shear capacity can be calculated using Eq. 7.9:

$$V_n = \min \begin{cases} 0.315A_n\sqrt{f'_m} & (a) \\ 2.07A_n & (b) \\ 0.621A_n + 0.45N & (c) \end{cases} \tag{7.9}$$

where A_n is the net cross sectional area of the wall.

PT-MWs having shear reinforcement can be considered as a reinforced masonry wall and the shear capacity can be calculated using Eq. 7.10:

$$V_n = 0.083 \left[4.0 - 1.75 \left(\frac{M}{Vl_w} \right) \right] A_n \sqrt{f'_m} + 0.25N + 0.5 \left(\frac{A_v}{s_h} \right) f_{yh} l_w \qquad (7.10)$$

where V_n is limited to $0.50 A_n \sqrt{f'_m}$ and $0.33 A_n \sqrt{f'_m}$ if $M/Vl_w \le 0.25$ and for $M/Vl_w \ge 1.00$, respectively, A_v and f_{yh} are the cross-sectional area and yield stress of the shear reinforcement, and s_h is the spacing of shear reinforcement.

In the flexural expression presented by MSJC (2013), different locations of PT bars are not considered. The equation was originally developed for out-of-plane loading in which the PT bars are usually located at the center of the wall, resulting in a single value of d. While acceptable for out-of-plane bending, for in-plane loading the equation is not able to account for the distribution of multiple PT bars along the length of the wall. Hence, Eqs. 7.7 and 7.8 need to be re-written as follows:

$$M_n = \sum f_{psi} A_{psi} \left(d_i - \frac{a}{2} \right) + \sum f_y A_{si} + N \left(\frac{l_w}{2} - \frac{a}{2} \right) \qquad (7.11)$$

$$a = \frac{\sum f_{psi} A_{psi} + f_y A_s + N}{0.8 f'_m b_w} \qquad (7.12)$$

where d_i is the distance between PT bar i and the extreme compression fiber.

For unbonded PT-MWs under in-plane loading the MSJC (2013) uses Eq. 7.13 to evaluate f_{ps}.

$$f_{ps} = f_{se} \qquad (7.13)$$

It is worth noting that Eq. 7.13 does not take into account the stress increment due to the elongation of PT bars. However, researchers reported that the current expression of MSJC (2013) provides a very conservative estimate of the strength (Wight 2006; Ryu et al. 2014).

Iterative Method

Recently, Hassanli et al. (2014b) developed the following design expression to predict the in-plane flexural strength of PT-MWs

$$f_{psi} = f_{sei} + \left(0.00055 L_w + 17.375 \frac{f_m}{f'_m} \right) \frac{E_{ps}}{L_{ps}} \left(\frac{d_i}{c} - 1 \right) \le f_{py} \qquad (7.14)$$

where

$$c = \frac{\sum f_{psi} A_{psi} + N}{\alpha \beta f'_m b} \qquad (7.15)$$

where L_{ps} is the unbonded length, E_{ps} is the elastic modulus of PT bar. α and β are the stress block parameters which are provided by different building codes (e.g. in MSJC (2013): $\alpha = \beta = 0.8$).

Simplified Method

Hassanli et al. (2014c) conducted a parametric study on PT-MWs and proposed the following non-iterative simplified equation to predict the stress developed in the PT bars at the peak in-plane strength.

$$f_{psi} = f_{sei} + \left(0.00055L_w + 17.375\frac{f_m}{f'_m}\right)\frac{E_{ps}}{L_p}\left(\frac{d_i f'_m}{2L_w f_m} - 1\right) \leq f_{py} \qquad (7.16)$$

While the total PT bar force of walls W1, W2, W3 and W4, obtained from the tests was 440.7, 554.6, 744.6 and 852.8 kN (Fig. 7.7), using iterative method (Eqs. 7.14 and 7.15) it is estimated to be equal to 443.7, 544.7, 581.7, and 861.9 kN and using simplified method (Eq. 7.16) to be equal to 443.2, 581.1, 671.6 and 847.4 kN, respectively. This shows that the both simplified and iterative methods could approximately predict the force developed in the PT bars.

The strength prediction obtained using the MSJC (2013) approach (No PT bar elongation); iterative method and simplified method are presented in Table 7.3. The predicted and tested strength are denoted by V_{EQN} and V_{EXP}, respectively. As shown in Table 7.3, the MSJC (2013) correctly predicted the modes of failure of the test specimens. The shear strengths were determined as 334 kN for walls W1 to W3 and 351 kN for wall W4, which are significantly higher than the predicted flexural strengths, implying that a flexural failure mode was expected as occurred during the experimental work. However, the MSJC (2013) was too conservative in predicting the flexural strengths. As presented in Table 7.3, the $V_{EQN}/V_{EXP(avg)}$ ratio using MSJC (2013) ranged from 0.41 to 0.71 with an average of 0.55.

Both the iterative and simplified approaches predicted the flexural strengths better than using MSJC (2013), presented in Table 7.3. The main reason for this significant difference is that the iterative and simplified approaches consider the elongation in the PT bars. As presented in Table 7.3, the $V_{EQN}/V_{EXP(avg)}$ ratio using the iterative method ranged from 0.80 to 0.99 with an average of 0.89, and using the simplified method ranged from 0.89 to 0.97 with an average of 0.95.

According to Table 7.1, as the number of PT bars increased from two in wall W1 to three in wall W2 and to four in wall W3 the initial to yield stress ratio in the PT bars (f_i/f_{py}) decreased from 0.63 in wall W1 to 0.42 in wall W2 and to 0.32 in wall W3. According to Table 7.3, based on MSJC (2013), the $V_{EQN}/V_{EXP(avg)}$ ratio decreased from 0.59 in wall W1, to 0.48 in wall W2 and to 0.41 in wall W3. This indicates that by ignoring the elongation of PT bar, MSJC (2013) provides more conservative results for the walls with a higher number of PT bars and lower values of f_i/f_{py} ratio. Comparing the results of the walls W3 and W4 reveals that MSJC (2013) provides a better prediction for wall W4. This can be explained by the

Table 7.3 Strength prediction of unbonded PT-MWs

Wall	V_{test}			V_{predic} (kN)			V_{predic}/V_{test} (avg)		
	Pull	Push	Average	MSJC (2013)	Iterative approach	Simplified approach	MSJC (2013)	Iterative approach	Simplified approach
W1	172.5	172.9	172.7	102.7	154.7	154.0	0.59	0.90	0.89
W2	209.0	217.7	213.4	102.7	189.1	206.2	0.48	0.89	0.97
W3	250.9	247.5	249.2	102.7	198.9	239.2	0.41	0.80	0.96
W4	266.6	240.6	253.6	179.5	251.8	244.6	0.71	0.99	0.96

different level of initial stress applied to the PT bars in the two walls. The initial stresses in the PT bars in wall W4 were twice those of the PT bars in wall W3, hence giving less capacity for the PT bars to develop stress before yielding.

7.8 Conclusion

MSJC (2013) requires limitations to ensure yielding of the post-tensioning steel prior to masonry compression failure in order to ensure a ductile behavior in unbonded PT-MWs. On the other hand, in the design of unbonded PT concrete walls according to the ACI criteria, the PT bars are designed to remain elastic to ensure self-centering behavior. The experimental research presented in this chapter suggests that both self-centering behavior and appropriate energy dissipation and a ductile response can be achieved, if an appropriate design philosophy is considered. Accordingly, the following design recommendations are proposed:

- The central PT bars should be designed to provide self-centering behavior, by ensuring that a portion of the PT force is retained in the central PT bars.
- The extreme tensile PT bars should be designed to yield prior to compression failure at the toe to provide ductility and energy dissipation.
- Shear reinforcement is required to maintain the integrity of the wall panel at high drifts, although it is not necessary for shear strength.

According to the results presented in this study, using MSJC (2013) approach to estimate the flexural strength of unbonded PT-MWs leads to very conservative results, due to ignoring the elongation of PT bars. This study demonstrated that the iterative and simplified equation developed recently to determine the in-plane strength of PT-MWs, could significantly improve the flexural strength prediction and that the theoretical approach based on geometric compatibility conditions could effectively predict the force-displacement response of unbonded PT-MWs.

References

ACI 318-08 (2008) Building code requirements for structural concrete and commentary. American Concrete Institute, Farmington Hills, MI

Adebar P, Ibrahim AM, Bryson M (2007) Test of high-rise core wall: effective stiffness for seismic analysis. ACI Struct J 104(5):549–559

ASTM C496 (2011) Standard test method for splitting tensile strength of cylindrical concrete specimens. ASTM International. West Conshohocken, PA

ASTM C1019 (2011) Standard test method for sampling and testing grout. ASTM International. West Conshohocken, PA

ASTM E2126 (2011) Standard test methods for cyclic (reversed) load test for shear resistance of vertical elements of the lateral force resisting systems for buildings. ASTM International. West Conshohocken, PA

ASTM C1314 (2012) Standard test method for compressive strength of masonry prisms. ASTM International. West Conshohocken, PA

ATSM C-140 (2013) Standard test methods for sampling and testing concrete masonry units and related units. ASTM International. West Conshohocken, PA

Banting B, El-Dakhakhni W (2014) Seismic design parameters for special masonry structural walls detailed with confined boundary elements. J Struct Eng 140(10):04014067

Ewing BD, Kowalsky MJ (2004) Compressive behavior of unconfined and confined clay brick masonry. J Struct Eng 130(4):650–661

FEMA-356 (2000) Pre-standard and commentary for the seismic rehabilitation of buildings. Federal Emergency Management Agency, Washington, D.C.

Hassanli R, Elgawady MA, Mills JE (2014a) Strength and seismic performance factors of post-tensioned masonry walls. J Struct Eng (in press)

Hassanli R, Elgawady MA, Mills JE (2014b) Flexural strength prediction of post-tensioned masonry walls. J Struct Eng (under review)

Hassanli R, Elgawady MA, Mills JE (2014c) Simplified approach to predict the flexural strength of unbonded post-tensioned masonry walls. J Struct Eng (under review)

Henry RS, Brooke NJ, Sritharan S, Ingham JM (2012) Defining concrete compressive strain in unbonded post-tensioned walls. ACI Struct J 109(1):101–111

Hose Y, Seible F (1999) Performance evaluation database for concrete bridge components and systems under simulated seismic loads. PEER report 1999/11. Berkley (USA): Pacific Earthquake Engineering Research Center, College of Engineering, University of California

IBC (2006) International building code. International Code Council, Inc. (formerly BOCA, ICBO and SBCCI), 4051, 60478-5795

Jacobsen LS (1930) Steady forced vibrations as influenced by damping. ASME Transactions 52 (1):169–181

Laursen PPT (2002) Seismic analysis and design of post-tensioned concrete masonry walls. Ph.D., dissertation, Department of Civil and Environmental Engineering, University of Auckland, Auckland, New Zealand

Laursen P, Ingham JM (2001) Structural testing of single-storey post-tensioned concrete masonry walls. Mason Soc J 19(1):69–82

Laursen P, Ingham JM (2004a) Structural testing of enhanced post-tensioned concrete masonry walls. ACI Struct J 101(6):852–862

Laursen P, Ingham JM (2004b) Structural testing of large-scale post-tensioned concrete masonry walls. J Struct Eng 130(10):1497–1505

Lissel SL, Shrive NG, Page AW (2000) Shear in plain, bed joint reinforced, and post-tensioned masonry. Can J Civ Eng 27(5):1021–1030

Masonry Standards Joint Committee (MSJC) (2013) Building code requirements for masonry structures. ACI 530/ASCE 5, TMS 402, American Concrete Institute, Detroit

Page A, Huizer A (1988) Racking behavior of pre-stressed and reinforced hollow masonry walls. Mason Int 2(3):97–102

Rosenboom OA (2002) Post-tensioned clay brick masonry walls for modular housing in seismic regions. M.S. thesis, North Carolina State University, Raleigh, NC, USA

Rosenboom OA, Kowalsky MJ (2004) Reversed in-plane cyclic behavior of post-tensioned clay brick masonry walls. J Struct Eng 130(5):787–798

Ryu D, Wijeyewickrema A, ElGawady M, Madurapperuma M (2014) Effects of tendon spacing on in-plane behavior of post-tensioned masonry walls. J Struct Eng 140(4):04013096

Schultz AE, Scolforo MJ (1991) Overview of pre-stressed masonry. Mason Soc J 10(1):6–21

Wight GD (2006) Seismic performance of a post-tensioned concrete masonry wall system. Ph.D., dissertation, Department of Civil and Environmental Engineering, University of Auckland, Auckland, New Zealand

Chapter 8
Summary and Conclusions

This chapter summarizes the outcomes of this study and proposes a number of recommendations for future research related to this subject.

8.1 Introduction

As stated in Chap. 1, the concept of pre-stressing of masonry, especially unbonded masonry is relatively new; hence, its behavior has not yet been well understood. This study has shown that the current procedures in the masonry design codes result in over-conservative predictions of the strength of unbonded post-tensioned masonry walls (PT-MWs). This was demonstrated by a combination of extensive analysis of experimental testing published to date, finite element modelling and analysis and an experimental study of four unbonded PT-MWs. The findings of the research have provided new insight into the performance of unbonded PT-MWs. Eight specific objectives of this study, which resulted in the development of eight research questions, were detailed in Chap. 1, as follows:

1. Study the accuracy of published design equations in predicting the flexural strength of unbonded PT-MWs based on a database of published experimental tests.
2. To provide finite element models to investigate the behavior of masonry prisms.
3. Predict the lateral load-displacement response of unbonded PT-MWs using an analytical procedure.
4. Develop design expressions and an analytical approach to predict the flexural strength and lateral force behavior of PT-MWs.
5. Provide design guidelines for unbonded PT-MWs.
6. Study the effect of the PT bar spacing and axial stress on the behavior of PT-MWs.

© Springer International Publishing AG, part of Springer Nature 2019
R. Hassanli, *Behavior of Unbounded Post-tensioned Masonry Walls*,
Springer Theses, https://doi.org/10.1007/978-3-319-93788-5_8

7. Provide experimental evidence to verify the accuracy of the proposed design methodology and guidelines.
8. Study the self-centering and energy dissipation behavior of unbonded PT-MWs.

8.2 Summary of the Research Undertaken

Following an extensive literature review, a database of published experimental results of laboratory testing of unbonded PT-MWs was collected and a comprehensive study was conducted to understand the seismic performance of such walls. The behavior of PT-MWs was examined according to the test results of 31 tested wall specimens (Chap. 3). MSJC (2013) standard has no procedure for estimating f_{ps} for unbonded PT-MWs. According to MSJC (2013), instead of a more accurate determination, f_{ps} for members with unbonded PT bars can conservatively be taken as f_{se}. The accuracy of this approach in predicting the strength of these walls was investigated based on the experimental database test results. The structural response parameters including ductility, response modification factor and displacement amplification factor were also studied.

A compressive prism and material modeling was provided in Chap. 4. The effects of the length, height and thickness on the compressive strength of concrete masonry prisms were evaluated in this chapter using calibrated finite element models. The accuracy of the height-to-thickness ratio correction factors provided by masonry codes was investigated.

Finite element models were constructed and calibrated (Chap. 5) in order to develop a design expression to predict the in-plane flexural strength of unbonded PT-MWs. To calibrate the material model, a finite element model of a masonry prism was developed and calibrated with experimental results. FEMs of six large scale PT-MWs were then developed and validated against experimental results. A parametric study was performed to investigate the effect of different parameters on the behavior and strength of PT-MWs. Multi-variate regression analysis was carried out to develop an equation to evaluate the rotation of PT-MWs at the peak strength. This equation was then incorporated into the flexural analysis of PT-MWs and an iterative method was developed to estimate the stress developed in the PT bars/tendons at flexural ultimate strength.

The accuracy of existing design expressions in predicting the flexural strength of PT-MWs was investigated and compared with the proposed expression in Chap. 6, based on available test results. Similarly, the expression estimating the wall rotation that was developed in Chap. 5 was implemented, in conjunction with a constitutive model, in an analytical procedure to predict the force-displacement response of PT-MWs. The procedure was validated against experimental test results of 11 PT-MWs. Using the validated procedure, a parametric study was performed to investigate the effect of different parameters on the compression zone length. Multi-variate regression analysis was then performed to develop an equation for the predicted compression zone length. In the obtained equation, the compression zone length was expressed as a function of wall length and axial stress ratio. This expression was incorporated into the previous iterative flexural analysis procedure in order to develop a simplified non-iterative expression to predict the stress developed in PT bars/tendons at ultimate flexural strength.

The behavior of PT-MWs was also investigated an experimental study. Four large-scale fully grouted unbonded concrete masonry walls were tested under reverse cyclic lateral in-plane load. The main variables considered in the experimental study were the bar spacing and the level of axial stress. The results of the experimental work were used to evaluate the design guidelines and expressions provided in Chaps. 5 and 6.

8.3 Summary of the Research Findings

The research findings have been summarized here in relation to the originally posed research objectives and questions:

Research Questions 1 and 7: What parameters influence the flexural and seismic behavior of unbonded PT-MWs?

As demonstrated in Chap. 3, relatively high values of R-factors were obtained for fully and partially grouted walls. It was recommended that R-factors of 2.5 and 3.0 for partially grouted and fully grouted PT-MWs, respectively, should be used. Using experimental results, it was shown that:

- The axial stress ratio has a prominent effect on the ductility. Based on the limited available data, to provide a ductile response, it is recommended to limit the axial stress ratio to a value of 0.15.
- Ungrouted prestressed walls displayed brittle behavior, characterized by a relatively small R-factor and ductility. These walls exhibited a limited displacement capacity and can be considered as ordinary plain masonry shear walls with minimal ductility. The recommended R-factor for these walls is 1.5.
- In almost all of the post-tensioned masonry walls tested so far that failed due to flexure, the PT bars have yielded. This results in an increased ductility, energy dissipation and a higher response modification factor. For these walls the supplemental mild steel is not required, as it does not increase the ductility of the system.

Research Questions 2: Is the masonry prism testing methods provided in the current design codes able to accurately reflect the actual strength of masonry?

Using a numerical finite element analysis to examine the influence of a range of factors on the compressive strength determined through the standard masonry prism tests, revealed that the current masonry prism testing method provided in MSJC (2013) is not able to accurately reflect the actual strength of masonry. Moreover, it was shown that

- The finite element analysis provided in Chap. 4, demonstrated that considering a height-to-thickness ratio of 2.0 in the ASTM C1314 (1314) standard results in a high over-prediction of the strength of the grouted concrete masonry prisms.
- The strength of the grouted prisms is not only functions of the height-to-thickness ratio but also the length of the prism. This needs to be addressed in masonry design codes in terms of strength correction factors.

Research Questions 3–5: How accurate are current design expressions and design codes in predicting a range of properties of PT-MWs?

Using experimental test results and finite element model results (Chaps. 5 and 6) it was shown that ignoring the elongation of PT bars/tendons in the flexural strength prediction, as is done in MSJC (2013), resulted in a too conservative strength prediction. Moreover, it was demonstrated that

- Using the shear expression provided in MSJC (2013), the shear strength of partially grouted and ungrouted post-tensioned walls was over-predicted by 12–86%. Hence, a revised shear strength equation is urgently needed.
- Using the strain compatibility method to estimate the strength of bonded post-tensioned walls resulted in an acceptable prediction.

Research Question 6 and 7: Is there an analytical procedure capable of predicting the flexural strength of PT-MWs? How accurate are the proposed design procedure and expressions?

To develop a new design approach to consider the elongation of the PT steel, a parametric study of finite element models of walls (Chap. 5) was conducted. As a result, the following iterative expression was proposed to estimate the stress developed in the PT bars/tendons at flexural ultimate strength.

$$f_{psi} = f_{sei} + \left(0.00055 L_w + 17.375 \frac{f_m}{f'_m} \right) \frac{E_{ps}}{L_{ps}} \left(\frac{d_i}{c} - 1 \right) \leq f_{py} \qquad (8.1)$$

Using experimental test results of 14 wall specimens from the test database demonstrated that the proposed approach leads to accurate and rational evaluation of the flexural strength of unbonded PT-MWs. It was concluded that the proposed expression could significantly improve the strength prediction of PT-MWs.

The force–displacement procedure considered in this study and presented in Chap. 6, was able to predict the lateral strength, stiffness and post-peak degradation behavior of the tested walls.

A parametric study was conducted using the validated analytical procedure. According to the results, the wall length and axial stress ratio were found to be the most significant factors affecting the compression zone length. Depending on the configuration of the wall, the compression zone length varies between 6.7 and 28% of the wall length. A multivariate regression analysis was performed to develop an empirical equation to estimate the compression zone length in unbonded PT-MWs.

This equation was then incorporated into the flexural analysis of PT-MWs and the following simplified non-iterative expression was developed to predict the stress developed in PT bars/tendons at ultimate flexural strength.

$$f_{ps\,i} = f_{se\,i} + \left(0.00055L_w + 17.375\frac{f_m}{f'_m}\right)\frac{E_{ps}}{L_p}\left(\frac{d_i f'_m}{2L_w f_m} - 1\right) \qquad (8.2)$$

The proposed simplified expression was then validated against experimental results and finite element model results. The strength of 14 tested PT-MWs of the database and two sets of FEMs were compared with the values calculated using the proposed simplified expression and the MSJC (2013) approach. According to the results, while disregarding the elongation of the PT bars in the unbonded PT-MW in MSJC (2013) resulted in a highly conservative strength prediction, the simplified expression could significantly improve the strength prediction.

To verify the accuracy of both the proposed iterative and simplified expression developed to predict the flexural strength, and also to further validate the presented analytical approach to predict the force-displacement behavior, an experimental study was conducted. A report on the experimental program was presented in Chap. 7. Four large-scale unbonded concrete masonry walls were tested under reverse cyclic lateral load. The main parameters of the experimental study were the PT bar spacing and the level of axial stress on the wall. The iterative method and simplified method could accurately predict the force-displacement response and approximately predict the total force developed in the PT bars of the test walls. Examination of the test data resulted in a number of significant conclusions regarding the behavior of unbonded PT-MWs. Of primary importance, the experimental research suggested that both self-centering behavior and appropriate energy dissipation response can be achieved in an unbonded PT-MW, if a proper design philosophy is considered. In the proposed design methodology, the central PT bars induce self-centring behavior to the system, and the extreme tensile PT bars provide the required ductility, energy dissipation and strength.

The accuracy of ignoring the elongation of PT bars in the MSJC (2013) approach and iterative and simplified design equations proposed in Chaps. 5 and 6, respectively, were examined to predict the flexural strength of the tested walls. Comparison of the experimental results and predictions from MSJC (2013) revealed that while ignoring the elongation of PT bars in the strength prediction resulted in a considerable underestimation of the strength, the two proposed design equations that included the elongation of the PT bars, could significantly improve the prediction. An investigation of the analytical approach developed in Chap. 6, verified the accuracy of the method to predict the force-displacement response of the tested walls. Finally, according to the results of the experimental work, shear reinforcement is required to maintain the integrity of the wall panel at high drifts, even though it is not necessary for shear strength.

8.4 Contribution to Knowledge

This research provides valuable design guidelines for unbonded PT-MWs. Two design methods, a simplified and iterative method, were developed and their accuracy was validated against test results. The presented analytical approach was able to successfully predict the force-displacement response.

This study also provided new knowledge on the self-centering response of unbonded PT-MWs. The results presented in this research suggested that both self-centering behavior and appropriate energy dissipation response can be achieved in an unbonded PT-MW, if a proper design philosophy is considered.

8.5 Recommendations for Future Work

The following research tasks were identified as being worthy of future research.

(1) While this research presented the first systematic approach to determine the seismic parameters for unbonded PT-MWs and determined the influence of different parameters on the seismic response of PT-MWs, there is an urgent need to enlarge the number of specimens in the database to confirm the conclusions from the current study.

(2) The seismic parameters presented in this study including ductility, response modification factor and displacement amplification factor, were focused on the component level and hence more research is required focusing on the response at the system level.

(3) The finite element analysis provided in Chap. 4, demonstrated that considering a height-to-thickness ratio of 2.0 in the ASTM C1314 (2003) standard results in a high over-prediction of the strength of the grouted concrete masonry prisms. Moreover, it was concluded that the strength of the grouted prisms is not only a function of the height-to-thickness ratio but also of the length of the prism. This implies that unlike the ASTM standard philosophy, the compressive strength obtained from testing of a full-length and half-length grouted prism may not be the same. The recommended strength correction provided here was developed based on a parametric study of finite element models. It is suggested that experimental tests should be conducted to validate the finite element model and the recommended correction factors.

(4) The analytical method that was developed in this thesis was successful in predicting the in-plane force-displacement response of unbonded PT-MWs. The method could potentially be applied to unbonded concrete PT walls. Further research is required to investigate the accuracy of the approach in predicting the lateral load behavior of concrete walls.

(5) The iterative method and simplified method developed in this research provided an appropriate estimation of the flexural strength of unbonded PT-MWs. Although this study proposed equations for unbonded PT-MWs, preliminary

investigation by the author revealed that the recommended equation could accurately predict the strength of post-tensioned concrete walls and hence may be applied to these members as well. More research is required to verify this fully.

(6) In-plane cyclic testing of unbonded PT-MWs was conducted in this study. There is limited experimental study to date on the dynamic behavior of PT-MWs. It is suggested that further experimental testing is required to determine the response of these walls under dynamic loads.

(7) The seismic behavior of structures comprising different post-tensioned elements needs to be studied experimentally and analytically. The interaction between the PT elements and the self-centering response of post-tensioned structures at system levels has not yet been well investigated.

References

ASTM C1314 (2003) Standard test method for compressive strength of masonry prisms. American Society for Testing and Materials, Pennsylvania, United States

Masonry Standards Joint Committee (MSJC) (2013) Building code requirements for masonry structures. ACI 530/ASCE 5, TMS 402, American Concrete Institute, Detroit

Appendix A
Calculation of T_c Values

This Appendix relates to work presented in Chap. 3 of the thesis.

The S_{1d}, S_{sd} and T_c values for 310 cities throughout the United States, sourced from H-18-8 (2013), with different seismicity and soil classes were calculated and presented in Table A.1.

Figures A.1, A.2, A.3 and A.4 indicate the frequency and the distribution of the T_c values of the cities belonging to the standard site classes. A normal distribution curve is also plotted on the figures for comparison. Figure C.5 indicates the cumulative frequency and the distribution of the T_c values of the cities belonging to the standard site classes. According to Figs. A.1, A.2, A.3, A.4 and A.5, the 95th percentile of T_c was determined to be equal to 0.46, 0.66, 0.69 and 1.01 for site classes A–B, C, D and E, respectively.

© Springer International Publishing AG, part of Springer Nature 2019
R. Hassanli, *Behavior of Unbounded Post-tensioned Masonry Walls*,
Springer Theses, https://doi.org/10.1007/978-3-319-93788-5

Table A.1 Seisemic parameters of 310 cities throughout the United States

Site	Center number	State	S_s	S_1	Seismicity	Site Class A			Site Class B		
						S_{ds}	S_{d1}	T_c	S_{ds}	S_{d1}	T_c
Abraham Lincoln	915	IL	0.19	0.07	L	0.10	0.04	0.36	0.13	0.05	0.36
Albany	528A8	NY	0.23	0.07	L	0.12	0.04	0.30	0.15	0.05	0.30
Albuquerque	501	NM	0.56	0.17	MH	0.30	0.09	0.30	0.37	0.11	0.30
Alexandria	502	LA	0.13	0.06	L	0.07	0.03	0.47	0.08	0.04	0.47
Alexandria	825	LA	0.13	0.06	L	0.07	0.03	0.48	0.09	0.04	0.48
Alexandria	826	VA	0.15	0.05	L	0.08	0.03	0.33	0.10	0.03	0.33
Allen Park	553A	MI	0.13	0.05	L	0.07	0.02	0.36	0.08	0.03	0.36
Alton	800	IL	0.48	0.15	MH	0.26	0.08	0.31	0.32	0.10	0.31
Altoona	503	PA	0.14	0.05	L	0.08	0.03	0.34	0.10	0.03	0.34
Amarillo	504	TX	0.18	0.04	L	0.09	0.02	0.25	0.12	0.03	0.25
American Lake	663A4	WA	1.19	0.42	H	0.64	0.22	0.35	0.80	0.28	0.35
Anchorage	463	AK	1.50	0.56	VH	0.80	0.30	0.37	1.00	0.37	0.37
Ann Arbor	506	MI	0.12	0.05	L	0.06	0.02	0.38	0.08	0.03	0.38
Annapolis	801	MD	0.16	0.05	L	0.08	0.03	0.32	0.10	0.03	0.32
Asheville	637	NC	0.39	0.11	MH	0.21	0.06	0.27	0.26	0.07	0.27
Aspinwall	646A4	PA	0.13	0.05	L	0.07	0.03	0.38	0.08	0.03	0.38
Atlanta	508	GA	0.23	0.09	L	0.12	0.05	0.37	0.15	0.06	0.37
Augusta	509	GA	0.37	0.11	MH	0.20	0.06	0.31	0.25	0.08	0.31
Augusta (Lenwood)	509A0	GA	0.38	0.12	MH	0.20	0.06	0.30	0.25	0.08	0.30
Balls Bluff	827	VA	0.16	0.05	L	0.09	0.03	0.32	0.11	0.03	0.32
Baltimore	512	MD	0.17	0.05	L	0.09	0.03	0.30	0.11	0.03	0.30
Baltimore	802	MD	0.17	0.05	L	0.09	0.03	0.30	0.11	0.03	0.30
Baltimore/Loch Raven	512GD	MD	0.17	0.05	L	0.09	0.03	0.30	0.11	0.03	0.30

(continued)

Table A.1 (continued)

Site	Center number	State	S$_s$	S$_1$	Seismicity	Site Class A			Site Class B		
						S$_{ds}$	S$_{d1}$	T$_c$	S$_{ds}$	S$_{d1}$	T$_c$
Barrancas	828	FL	0.10	0.05	L	0.05	0.03	0.48	0.07	0.03	0.48
Batavia	528A4	NY	0.27	0.06	ML	0.14	0.03	0.23	0.18	0.04	0.23
Bath	803	NY	0.17	0.05	L	0.09	0.03	0.32	0.11	0.04	0.32
Bath	528A6	NY	0.17	0.05	L	0.09	0.03	0.32	0.11	0.04	0.32
Baton Rouge	829	LA	0.12	0.05	L	0.06	0.03	0.44	0.08	0.04	0.44
Battle Creek	515	MI	0.11	0.05	L	0.06	0.03	0.42	0.07	0.03	0.42
Bay Pines	516	FL	0.08	0.03	L	0.04	0.02	0.41	0.05	0.02	0.41
Bay Pines	830	FL	0.08	0.03	L	0.04	0.02	0.41	0.05	0.02	0.41
Beaufort	831	SC	0.69	0.18	MH	0.37	0.10	0.26	0.46	0.12	0.26
Beckley	517	WV	0.26	0.08	ML	0.14	0.04	0.29	0.18	0.05	0.29
Bedford	518	MA	0.29	0.07	ML	0.15	0.04	0.24	0.19	0.05	0.24
Beverly	804	NJ	0.27	0.06	ML	0.15	0.03	0.22	0.18	0.04	0.22
Big Spring	519	TX	0.11	0.03	L	0.06	0.02	0.28	0.07	0.02	0.28
Biloxi	520	MS	0.12	0.05	L	0.06	0.03	0.44	0.08	0.03	0.44
Biloxi	832	MS	0.12	0.05	L	0.06	0.03	0.44	0.08	0.03	0.44
Birmingham	521	AL	0.30	0.10	ML	0.16	0.05	0.32	0.20	0.06	0.32
Black Hills	884	SD	0.15	0.04	L	0.08	0.02	0.28	0.10	0.03	0.28
Boise	531	ID	0.31	0.11	ML	0.17	0.06	0.34	0.21	0.07	0.34
Bonham	549A4	TX	0.16	0.06	L	0.09	0.03	0.39	0.11	0.04	0.39
Boston	523	MA	0.27	0.07	ML	0.15	0.04	0.25	0.18	0.04	0.25
Brevard	673GA	FL	0.08	0.04	L	0.04	0.02	0.42	0.06	0.02	0.42
Brockton	523A5	MA	0.25	0.06	ML	0.13	0.03	0.25	0.17	0.04	0.25
Bronx	526	NY	0.36	0.07	MH	0.19	0.04	0.19	0.24	0.05	0.19

(continued)

Table A.1 (continued)

Site	Center number	State	S_s	S_1	Seismicity	Site Class A			Site Class B		
						S_{ds}	S_{d1}	T_c	S_{ds}	S_{d1}	T_c
Brooklyn	630A4	NY	0.35	0.07	MH	0.19	0.04	0.20	0.24	0.05	0.20
Buffalo	528	NY	0.28	0.06	ML	0.15	0.03	0.21	0.19	0.04	0.21
Butler	529	PA	0.13	0.05	L	0.07	0.03	0.38	0.09	0.03	0.38
Calverton	805	NY	0.21	0.06	L	0.11	0.03	0.27	0.14	0.04	0.27
Camp Butler	806	IL	0.27	0.11	L	0.14	0.06	0.39	0.18	0.07	0.39
Camp Nelson	833	KY	0.23	0.09	ML	0.12	0.05	0.40	0.15	0.06	0.40
Canandaigua	528A5	NY	0.19	0.06	L	0.10	0.03	0.30	0.12	0.04	0.30
Castle Point	620A4	NY	0.28	0.07	ML	0.15	0.04	0.24	0.19	0.04	0.24
Cave Hill	834	KY	0.25	0.10	ML	0.13	0.05	0.42	0.16	0.07	0.42
Charleston	534	SC	1.44	0.36	VH	0.77	0.19	0.25	0.96	0.24	0.25
Chattanooga	835	TN	0.47	0.12	MH	0.25	0.06	0.24	0.32	0.08	0.24
Cheyenne	442	WY	0.19	0.05	L	0.10	0.03	0.28	0.13	0.04	0.28
Chicago (Lakeside)	537GD	IL	0.16	0.06	L	0.09	0.03	0.37	0.11	0.04	0.37
Chicago (Westside)	537	IL	0.17	0.06	L	0.09	0.03	0.36	0.11	0.04	0.36
Chillicothe	538	OH	0.16	0.06	L	0.08	0.03	0.41	0.10	0.04	0.41
Cincinnati	539	OH	0.18	0.08	L	0.09	0.04	0.43	0.12	0.05	0.43
City Point	836	VA	0.19	0.06	L	0.10	0.03	0.31	0.12	0.04	0.31
Clarksburg	540	WV	0.19	0.07	L	0.10	0.04	0.36	0.12	0.05	0.36
Cleveland/Brecksville	541A0	OH	0.20	0.05	L	0.11	0.03	0.26	0.13	0.03	0.26
Cleveland/Wade Park	541	OH	0.20	0.05	L	0.11	0.03	0.26	0.13	0.03	0.26
Coatesville	542	PA	0.27	0.06	ML	0.15	0.03	0.22	0.18	0.04	0.22
Cold Harbor	837	VA	0.21	0.06	L	0.11	0.03	0.28	0.14	0.04	0.28
Columbia	589A4	MO	0.20	0.09	L	0.11	0.05	0.44	0.13	0.06	0.44

(continued)

Table A.1 (continued)

Site	Center number	State	S_s	S_1	Seismicity	Site Class A			Site Class B		
						S_{ds}	S_{d1}	T_c	S_{ds}	S_{d1}	T_c
Columbia	544	SC	0.57	0.15	MH	0.31	0.08	0.27	0.38	0.10	0.27
Corinth	838	MS	0.50	0.17	MH	0.27	0.09	0.34	0.33	0.11	0.34
Crown Hill	807	IN	0.19	0.08	L	0.10	0.04	0.44	0.13	0.06	0.44
Culpeper	839	VA	0.19	0.06	L	0.10	0.03	0.30	0.13	0.04	0.30
Cypress Hills	808	NY	0.36	0.07	MH	0.19	0.04	0.19	0.24	0.05	0.19
Dallas	549	TX	0.11	0.05	L	0.06	0.03	0.43	0.08	0.03	0.43
Dallas/Fort Worth	916	TX	0.12	0.05	L	0.06	0.03	0.43	0.08	0.03	0.43
Danville	550	IL	0.22	0.09	L	0.12	0.05	0.41	0.15	0.06	0.41
Danville	809	IL	0.22	0.09	L	0.12	0.05	0.41	0.15	0.06	0.41
Danville	840	KY	0.22	0.09	L	0.12	0.05	0.42	0.15	0.06	0.42
Danville	841	VA	0.20	0.07	L	0.10	0.04	0.38	0.13	0.05	0.38
Dayton	552	OH	0.19	0.07	L	0.10	0.04	0.37	0.12	0.05	0.37
Dayton	810	OH	0.21	0.07	L	0.11	0.04	0.34	0.14	0.05	0.34
Denver	554	CO	0.21	0.06	L	0.11	0.03	0.26	0.14	0.04	0.26
Des Moines	636A6	IA	0.08	0.04	L	0.04	0.02	0.57	0.05	0.03	0.57
Detroit	553	MI	0.12	0.05	L	0.06	0.02	0.37	0.08	0.03	0.37
Dublin	557	GA	0.22	0.08	L	0.11	0.04	0.39	0.14	0.06	0.39
Durham	558	NC	0.20	0.08	L	0.11	0.04	0.39	0.13	0.05	0.39
Eagle Point	906	OR	0.58	0.26	MH	0.31	0.14	0.44	0.39	0.17	0.44
East Orange	561	NJ	0.36	0.07	MH	0.19	0.04	0.20	0.24	0.05	0.20
El Paso	756	TX	0.33	0.11	ML	0.18	0.06	0.32	0.22	0.07	0.32
Erie	562	PA	0.16	0.05	L	0.09	0.03	0.30	0.11	0.03	0.30
Fargo	437	ND	0.07	0.02	L	0.04	0.01	0.28	0.05	0.01	0.28

(continued)

Table A.1 (continued)

Site	Center number	State	S_s	S_1	Seismicity	Site Class A			Site Class B		
						S_{ds}	S_{d1}	T_c	S_{ds}	S_{d1}	T_c
Fayetteville	564	AR	0.21	0.09	L	0.11	0.05	0.44	0.14	0.06	0.44
Fayetteville	565	NC	0.30	0.10	ML	0.16	0.05	0.34	0.20	0.07	0.34
Fayetteville	842	AR	0.21	0.09	L	0.11	0.05	0.44	0.14	0.06	0.44
Finn's Point	811	NJ	0.23	0.06	L	0.12	0.03	0.24	0.15	0.04	0.24
Florence	843	SC	0.73	0.20	MH	0.39	0.11	0.27	0.49	0.13	0.27
Florida	911	FL	0.09	0.04	L	0.05	0.02	0.42	0.06	0.03	0.42
Fort Bayard	885	NM	0.27	0.08	ML	0.15	0.04	0.30	0.18	0.05	0.30
Fort Bliss	886	TX	0.34	0.11	MH	0.18	0.06	0.32	0.22	0.07	0.32
Fort Custer	909	MI	0.11	0.05	L	0.06	0.03	0.43	0.07	0.03	0.43
Fort Gibson	844	OK	0.19	0.08	L	0.10	0.04	0.40	0.12	0.05	0.40
Fort Harrison	436	MT	0.75	0.22	MH	0.40	0.12	0.30	0.50	0.15	0.30
Fort Harrison	845	VA	0.23	0.06	L	0.12	0.03	0.27	0.15	0.04	0.27
Fort Howard	512GF	MD	0.17	0.05	L	0.09	0.03	0.30	0.11	0.03	0.30
Fort Leavenworth	887	KS	0.13	0.06	L	0.07	0.03	0.43	0.09	0.04	0.43
Fort Logan	888	CO	0.22	0.06	L	0.12	0.03	0.26	0.15	0.04	0.26
Fort Lyon	567	CO	0.17	0.05	L	0.09	0.03	0.30	0.11	0.03	0.30
Fort Lyon	889	CO	0.17	0.05	L	0.09	0.03	0.30	0.11	0.03	0.30
Fort McPherson	890	NE	0.09	0.03	L	0.05	0.02	0.35	0.06	0.02	0.35
Fort Meade	568	SD	0.21	0.05	L	0.11	0.03	0.25	0.14	0.03	0.25
Fort Meade	891	SD	0.21	0.05	L	0.11	0.03	0.25	0.14	0.03	0.25
Fort Mitchell	908	AL	0.14	0.07	L	0.08	0.04	0.46	0.09	0.04	0.46
Fort Richardson	910	AK	1.50	0.56	VH	0.80	0.30	0.37	1.00	0.37	0.37
Fort Rosecrans	892	CA	1.57	0.61	VH	0.84	0.33	0.39	1.05	0.41	0.39

(continued)

Table A.1 (continued)

Site	Center number	State	S_s	S_1	Seismicity	Site Class A			Site Class B		
						S_{ds}	S_{d1}	T_c	S_{ds}	S_{d1}	T_c
Fort Sam Houston	846	TX	0.11	0.03	L	0.06	0.02	0.28	0.07	0.02	0.28
Fort Scott	893	KS	0.13	0.07	L	0.07	0.04	0.52	0.09	0.04	0.52
Fort Sill	920	OK	0.37	0.09	MH	0.20	0.05	0.23	0.25	0.06	0.23
Fort Smith	847	AR	0.21	0.09	L	0.11	0.05	0.42	0.14	0.06	0.42
Fort Snelling	894	MN	0.06	0.03	L	0.03	0.01	0.44	0.04	0.02	0.44
Fort Thomas	539A	OH	0.15	0.06	L	0.08	0.03	0.39	0.10	0.04	0.39
Fort Wayne	610A4	IN	0.15	0.06	L	0.08	0.03	0.39	0.10	0.04	0.39
Fresno	570	CA	0.50	0.22	MH	0.27	0.12	0.44	0.33	0.15	0.44
Gainesville	573	FL	0.11	0.05	L	0.06	0.03	0.45	0.07	0.03	0.45
Glendale	848	VA	0.23	0.06	L	0.12	0.03	0.27	0.15	0.04	0.27
Golden Gate	895	CA	2.22	1.27	VH	1.18	0.68	0.57	1.48	0.84	0.57
Grafton	812	WV	0.14	0.05	L	0.07	0.03	0.39	0.09	0.04	0.39
Grand Island	636A4	NE	0.13	0.04	L	0.07	0.02	0.30	0.09	0.03	0.30
Grand Junction	575	CO	0.29	0.07	ML	0.15	0.04	0.23	0.19	0.04	0.23
Gulfport	520A0	MS	0.12	0.05	L	0.06	0.03	0.44	0.08	0.03	0.44
Hampton	590	VA	0.12	0.05	L	0.06	0.03	0.40	0.08	0.03	0.40
Hampton	849	VA	0.12	0.05	L	0.07	0.03	0.40	0.08	0.03	0.40
Hampton (VAMC)	850	VA	0.12	0.05	L	0.07	0.03	0.40	0.08	0.03	0.40
Hines	578	IL	0.17	0.06	L	0.09	0.03	0.34	0.12	0.04	0.34
Hines VBA	201	IL	0.17	0.06	L	0.09	0.03	0.34	0.12	0.04	0.34
Honolulu**	459	HI	0.61	0.18	MH	0.33	0.09	0.29	0.41	0.12	0.29
Hot Springs	896	SD	0.21	0.05	L	0.11	0.03	0.24	0.14	0.03	0.24
Hot Springs	568A4	SD	0.21	0.05	L	0.11	0.03	0.24	0.14	0.03	0.24

(continued)

Table A.1 (continued)

Site	Center number	State	S_s	S_1	Seismicity	Site Class A			Site Class B		
						S_{ds}	S_{d1}	T_c	S_{ds}	S_{d1}	T_c
Houston	580	TX	0.09	0.04	L	0.05	0.02	0.41	0.06	0.02	0.41
Houston	851	TX	0.09	0.04	L	0.05	0.02	0.41	0.06	0.02	0.41
Houston VBA	362	TX	0.09	0.04	L	0.05	0.02	0.41	0.06	0.02	0.41
Huntington	581	WV	0.19	0.07	L	0.10	0.04	0.37	0.13	0.05	0.37
Indianapolis	583	IN	0.19	0.08	L	0.10	0.04	0.43	0.13	0.06	0.43
Indianapolis (CS Rd)	583A4	IN	0.19	0.08	L	0.10	0.04	0.43	0.13	0.06	0.43
Indiantown Gap	813	PA	0.22	0.06	L	0.12	0.03	0.26	0.15	0.04	0.26
Iowa City	636A8	IA	0.10	0.05	L	0.05	0.03	0.52	0.07	0.03	0.52
Iron Mountain	585	MI	0.06	0.03	L	0.03	0.01	0.45	0.04	0.02	0.45
Jackson	586	MS	0.19	0.09	L	0.10	0.05	0.44	0.13	0.06	0.44
Jackson VBA	323	MS	0.19	0.09	L	0.10	0.05	0.44	0.13	0.06	0.44
Jefferson Barracks	852	MO	0.58	0.17	MH	0.31	0.09	0.29	0.39	0.11	0.29
Jefferson City	853	MO	0.24	0.10	ML	0.13	0.05	0.42	0.16	0.07	0.42
Kansas City	589	MO	0.13	0.06	L	0.07	0.03	0.46	0.08	0.04	0.46
Keokuk	814	IA	0.15	0.07	L	0.08	0.04	0.49	0.10	0.05	0.49
Kerrville	854	TX	0.07	0.03	L	0.04	0.01	0.35	0.05	0.02	0.35
Kerrville	671A4	TX	0.07	0.03	L	0.04	0.01	0.35	0.05	0.02	0.35
Knoxville	636A7	IA	0.08	0.05	L	0.04	0.03	0.58	0.06	0.03	0.58
Knoxville	855	TN	0.52	0.12	MH	0.28	0.06	0.23	0.35	0.08	0.23
Lake City	573A4	FL	0.12	0.05	L	0.07	0.03	0.44	0.08	0.04	0.44
Las Vegas	593	NV	0.55	0.17	MH	0.29	0.09	0.31	0.37	0.11	0.31
Leavenworth	897	KS	0.13	0.06	L	0.07	0.03	0.43	0.09	0.04	0.43
Leavenworth	589A6	KS	0.13	0.06	L	0.07	0.03	0.43	0.09	0.04	0.43

(continued)

Table A.1 (continued)

Site	Center number	State	S_s	S_1	Seismicity	Site Class A			Site Class B		
						S_{ds}	S_{d1}	T_c	S_{ds}	S_{d1}	T_c
Lebanon	856	KY	0.23	0.10	L	0.12	0.05	0.44	0.15	0.07	0.44
Lebanon	595	PA	0.23	0.06	L	0.12	0.03	0.25	0.15	0.04	0.25
Lexington	857	KY	0.22	0.09	L	0.12	0.05	0.39	0.15	0.06	0.39
Lexington (CD)	596A4	KY	0.23	0.09	L	0.12	0.05	0.38	0.15	0.06	0.38
Lexington (LD)	596	KY	0.23	0.09	L	0.12	0.05	0.38	0.15	0.06	0.38
Lincoln	636A5	NE	0.18	0.05	L	0.09	0.02	0.26	0.12	0.03	0.26
Little Rock	598	AR	0.49	0.16	MH	0.26	0.09	0.32	0.33	0.11	0.32
Little Rock	858	AR	0.51	0.16	MH	0.27	0.09	0.32	0.34	0.11	0.32
Livermore	640A4	CA	1.59	0.60	VH	0.85	0.32	0.38	1.06	0.40	0.38
Loma Linda	605	CA	1.76	0.61	VH	0.94	0.33	0.35	1.17	0.41	0.35
Long Beach	600	CA	2.02	0.85	VH	1.08	0.45	0.42	1.35	0.57	0.42
Long Island	815	NY	0.29	0.06	ML	0.16	0.03	0.22	0.20	0.04	0.22
Los Angeles	898	CA	1.66	0.59	VH	0.88	0.31	0.36	1.10	0.39	0.36
Los Angeles	691GE	CA	2.23	0.77	VH	1.19	0.41	0.34	1.49	0.51	0.34
Loudon Park	816	MD	0.17	0.05	L	0.09	0.03	0.30	0.11	0.03	0.30
Louisville	603	KY	0.25	0.10	ML	0.13	0.05	0.41	0.16	0.07	0.41
Lyons	561A4	NJ	0.35	0.07	ML	0.19	0.04	0.20	0.23	0.05	0.20
Madison	607	WI	0.10	0.04	L	0.06	0.02	0.42	0.07	0.03	0.42
Manchester	608	NH	0.35	0.08	MH	0.19	0.04	0.23	0.23	0.05	0.23
Marietta	859	GA	0.25	0.09	ML	0.13	0.05	0.35	0.17	0.06	0.35
Marion	657A5	IL	1.12	0.31	H	0.60	0.16	0.27	0.75	0.20	0.27
Marion	610	IN	0.15	0.07	L	0.08	0.04	0.44	0.10	0.04	0.44
Marion	817	IN	0.15	0.07	L	0.08	0.04	0.44	0.10	0.04	0.44

(continued)

Table A.1 (continued)

Site	Center number	State	S_s	S_1	Seismicity	Site Class A			Site Class B		
						S_{ds}	S_{d1}	T_c	S_{ds}	S_{d1}	T_c
Martin	674A5	TX	0.09	0.04	L	0.05	0.02	0.44	0.06	0.03	0.44
Martinez/NCSC	612	CA	1.58	0.60	VH	0.84	0.32	0.38	1.05	0.40	0.38
Martinsburg	613	WV	0.17	0.05	L	0.09	0.03	0.31	0.11	0.03	0.31
Massachusetts	818	MA	0.21	0.06	L	0.11	0.03	0.27	0.14	0.04	0.27
McClellan	612GH	CA	0.49	0.22	MH	0.26	0.12	0.45	0.32	0.15	0.45
Memphis	614	TN	1.29	0.35	VH	0.69	0.19	0.27	0.86	0.24	0.27
Memphis	860	TN	1.29	0.35	VH	0.69	0.19	0.27	0.86	0.24	0.27
Menlo Park	640A0	CA	1.79	0.79	VH	0.96	0.42	0.44	1.19	0.53	0.44
Miami	546	FL	0.05	0.02	L	0.03	0.01	0.37	0.03	0.01	0.37
Miles City	436GJ	MT	0.10	0.03	L	0.05	0.02	0.34	0.07	0.02	0.34
Mill Springs	861	KY	0.23	0.10	L	0.12	0.05	0.41	0.16	0.06	0.41
Milwaukee (Wood)	695	WI	0.11	0.05	L	0.06	0.02	0.42	0.07	0.03	0.42
Minneapolis	618	MN	0.06	0.03	L	0.03	0.01	0.45	0.04	0.02	0.45
Mobile	862	AL	0.12	0.05	L	0.06	0.03	0.45	0.08	0.04	0.45
Montgomery	619	AL	0.15	0.07	L	0.08	0.04	0.45	0.10	0.05	0.45
Montgomery VBA	322	AL	0.15	0.07	L	0.08	0.04	0.45	0.10	0.05	0.45
Montrose	620	NY	0.33	0.07	ML	0.18	0.04	0.21	0.22	0.05	0.21
Mound City	863	IL	3.39	1.31	VH	1.81	0.70	0.39	2.26	0.88	0.39
Mountain Home	621	TN	0.39	0.10	MH	0.21	0.05	0.26	0.26	0.07	0.26
Mountain Home	864	TN	0.39	0.10	MH	0.21	0.05	0.26	0.26	0.07	0.26
Murfreesboro	626A4	TN	0.29	0.12	ML	0.15	0.06	0.40	0.19	0.08	0.40
Muskogee	623	OK	0.19	0.07	L	0.10	0.04	0.40	0.12	0.05	0.40
Nashville	626	TN	0.35	0.13	ML	0.18	0.07	0.39	0.23	0.09	0.39

(continued)

Table A.1 (continued)

Site	Center number	State	S_s	S_1	Seismicity	Site Class A			Site Class B		
						S_{ds}	S_{d1}	T_c	S_{ds}	S_{d1}	T_c
Nashville	865	TN	0.35	0.13	ML	0.18	0.07	0.39	0.23	0.09	0.39
Natchez	866	MS	0.14	0.07	L	0.08	0.04	0.48	0.09	0.04	0.48
NCA Operations Support	786	VA	0.19	0.07	L	0.10	0.04	0.35	0.13	0.04	0.35
New Albany	867	IN	0.25	0.10	ML	0.13	0.05	0.41	0.17	0.07	0.41
New Bern	868	NC	0.16	0.07	L	0.09	0.03	0.40	0.11	0.04	0.40
New Orleans	629	LA	0.11	0.05	L	0.06	0.03	0.44	0.07	0.03	0.44
New York	630	NY	0.36	0.07	MH	0.19	0.04	0.19	0.24	0.05	0.19
Newington	689A4	CT	0.24	0.06	L	0.13	0.03	0.26	0.16	0.04	0.26
NMCA	914	AZ	0.18	0.06	L	0.09	0.03	0.34	0.12	0.04	0.34
NMCP	899	HI	0.61	0.18	MH	0.33	0.09	0.29	0.41	0.12	0.29
North Chicago	556	IL	0.14	0.05	L	0.08	0.03	0.38	0.09	0.04	0.38
North Little Rock	598A0	AR	0.51	0.17	MH	0.27	0.09	0.32	0.34	0.11	0.32
Northampton	631	MA	0.22	0.07	LL	0.12	0.04	0.29	0.15	0.04	0.29
Northport	632	NY	0.29	0.06	ML	0.15	0.03	0.22	0.19	0.04	0.22
Oklahoma City	635	OK	0.34	0.07	ML	0.18	0.04	0.22	0.22	0.05	0.22
Omaha	636	NE	0.12	0.04	L	0.07	0.02	0.34	0.08	0.03	0.34
Orlando	673BY	FL	0.10	0.04	L	0.05	0.02	0.40	0.06	0.03	0.40
Palo Alto	640	CA	1.96	0.83	VH	1.04	0.44	0.42	1.31	0.55	0.42
Perry Point	512A5	MD	0.22	0.05	L	0.12	0.03	0.25	0.14	0.04	0.25
Philadelphia	642	PA	0.27	0.06	ML	0.14	0.03	0.22	0.18	0.04	0.22
Philadelphia	819	PA	0.28	0.06	ML	0.15	0.03	0.22	0.19	0.04	0.22
Phoenix	644	AZ	0.18	0.06	L	0.10	0.03	0.34	0.12	0.04	0.34
Pittsburgh (HD)	646A5	PA	0.13	0.05	L	0.07	0.03	0.39	0.08	0.03	0.39

(continued)

Table A.1 (continued)

Site	Center number	State	S_s	S_1	Seismicity	Site Class A			Site Class B		
						S_{ds}	S_{d1}	T_c	S_{ds}	S_{d1}	T_c
Pittsburgh (UD)	646	PA	0.13	0.05	L	0.07	0.03	0.38	0.08	0.03	0.38
Poplar Bluff	657A4	MO	1.10	0.30	H	0.59	0.16	0.28	0.73	0.20	0.28
Port Hudson	870	LA	0.12	0.06	L	0.07	0.03	0.45	0.08	0.04	0.45
Portland	648	OR	0.98	0.35	H	0.52	0.18	0.35	0.66	0.23	0.35
Prescott	649	AZ	0.34	0.10	ML	0.18	0.05	0.29	0.23	0.07	0.29
Prescott	900	AZ	0.35	0.10	MH	0.19	0.05	0.29	0.23	0.07	0.29
Providence	650	RI	0.23	0.06	L	0.12	0.03	0.26	0.16	0.04	0.26
Quantico	872	VA	0.16	0.05	L	0.09	0.03	0.32	0.11	0.03	0.32
Quincy	820	IL	0.18	0.08	L	0.10	0.04	0.45	0.12	0.05	0.45
Raleigh	873	NC	0.20	0.08	L	0.11	0.04	0.39	0.13	0.05	0.39
Reno	654	NV	1.50	0.60	VH	0.80	0.32	0.40	1.00	0.40	0.40
Richmond	652	VA	0.23	0.06	L	0.12	0.03	0.27	0.15	0.04	0.27
Richmond	874	VA	0.23	0.06	L	0.12	0.03	0.27	0.15	0.04	0.27
Riverside	901	CA	1.50	0.60	VH	0.80	0.32	0.40	1.00	0.40	0.40
Rock Island	821	IL	0.13	0.06	L	0.07	0.03	0.46	0.09	0.04	0.46
Roseburg	653	OR	0.83	0.42	VH	0.44	0.23	0.51	0.55	0.28	0.51
Roseburg	902	OR	0.83	0.42	H	0.44	0.23	0.51	0.55	0.28	0.51
Sacramento NCHCS	612A4	CA	0.46	0.21	MH	0.25	0.11	0.46	0.31	0.14	0.46
Saginaw	655	MI	0.08	0.04	L	0.04	0.02	0.46	0.05	0.02	0.46
Salem	658	VA	0.26	0.08	ML	0.14	0.04	0.29	0.18	0.05	0.29
Salisbury	659	NC	0.26	0.09	ML	0.14	0.05	0.36	0.17	0.06	0.36
Salisbury	876	NC	0.26	0.09	ML	0.14	0.05	0.36	0.17	0.06	0.36
Salt Lake City	660	UT	1.58	0.63	VH	0.84	0.33	0.40	1.05	0.42	0.40

(continued)

Table A.1 (continued)

Site	Center number	State	S_s	S_1	Seismicity	Site Class A			Site Class B		
						S_{ds}	S_{d1}	T_c	S_{ds}	S_{d1}	T_c
San Antonio	671	TX	0.11	0.03	L	0.06	0.02	0.30	0.07	0.02	0.30
San Antonio	877	TX	0.11	0.03	L	0.06	0.02	0.28	0.07	0.02	0.28
San Diego	664	CA	1.56	0.60	VH	0.83	0.32	0.39	1.04	0.40	0.39
San Francisco	662	CA	1.76	0.90	VH	0.94	0.48	0.51	1.17	0.60	0.51
San Francisco	903	CA	1.50	0.67	VH	0.80	0.36	0.45	1.00	0.45	0.45
San Joaquin Valley	913	CA	1.83	0.60	VH	0.97	0.32	0.33	1.22	0.40	0.33
San Juan	672	PR	0.90	0.31	H	0.48	0.17	0.35	0.60	0.21	0.35
Santa Fe	904	NM	0.48	0.16	MH	0.26	0.08	0.32	0.32	0.10	0.32
Saratoga	917	NY	0.25	0.07	ML	0.13	0.04	0.29	0.17	0.05	0.29
Seattle	663	WA	1.55	0.53	VH	0.83	0.28	0.34	1.03	0.36	0.34
Sepulveda	691A4	CA	2.04	0.73	VH	1.09	0.39	0.36	1.36	0.48	0.36
Seven Pines	878	VA	0.20	0.06	L	0.11	0.03	0.29	0.13	0.04	0.29
Sheridan	666	WY	0.27	0.06	ML	0.14	0.03	0.22	0.18	0.04	0.22
Shreveport	667	LA	0.15	0.07	L	0.08	0.04	0.45	0.10	0.05	0.45
Sioux Falls	438	SD	0.11	0.03	L	0.06	0.02	0.31	0.07	0.02	0.31
Sitka	905	AK	0.97	0.50	H	0.51	0.27	0.52	0.64	0.33	0.52
Somerville AMS	796	NJ	0.33	0.07	ML	0.17	0.04	0.21	0.22	0.04	0.21
Spokane	668	WA	0.40	0.11	MH	0.22	0.06	0.28	0.27	0.08	0.28
Springfield	879	MO	0.22	0.10	L	0.12	0.05	0.43	0.15	0.06	0.43
St. Albans	630A5	NY	0.34	0.07	ML	0.18	0.04	0.20	0.23	0.04	0.20
St. Augustine	875	FL	0.13	0.05	L	0.07	0.03	0.42	0.08	0.04	0.42
St. Cloud	656	MN	0.08	0.02	L	0.04	0.01	0.28	0.05	0.01	0.28
St. Louis (JB)	657A0	MO	0.60	0.17	MH	0.32	0.09	0.29	0.40	0.11	0.29

(continued)

Table A.1 (continued)

Site	Center number	State	S_s	S_1	Seismicity	Site Class A			Site Class B		
						S_{ds}	S_{d1}	T_c	S_{ds}	S_{d1}	T_c
St. Louis (JC)	657	MO	0.60	0.17	MH	0.32	0.09	0.29	0.40	0.11	0.29
St. Petersburg VBA	317	FL	0.08	0.03	L	0.04	0.02	0.41	0.05	0.02	0.41
Staunton	880	VA	0.21	0.06	L	0.11	0.03	0.30	0.14	0.04	0.30
Syracuse	528A7	NY	0.18	0.06	L	0.10	0.03	0.34	0.12	0.04	0.34
Tahoma	919	WA	1.28	0.44	VH	0.68	0.23	0.34	0.86	0.29	0.34
Tampa	673	FL	0.08	0.03	L	0.04	0.02	0.42	0.05	0.02	0.42
Temple	674	TX	0.08	0.04	L	0.04	0.02	0.45	0.05	0.02	0.45
Togus	402	ME	0.29	0.08	ML	0.16	0.04	0.26	0.19	0.05	0.26
Togus	822	ME	0.29	0.08	ML	0.16	0.04	0.26	0.19	0.05	0.26
Tomah	676	WI	0.07	0.03	L	0.04	0.02	0.51	0.04	0.02	0.51
Topeka	589A5	KS	0.16	0.05	L	0.08	0.03	0.34	0.10	0.04	0.34
Tucson	678	AZ	0.29	0.08	ML	0.15	0.04	0.28	0.19	0.05	0.28
Tuscaloosa	679	AL	0.27	0.09	ML	0.14	0.05	0.35	0.18	0.06	0.35
Tuskegee	619A4	AL	0.15	0.07	L	0.08	0.04	0.45	0.10	0.05	0.45
Vancouver	648A4	WA	0.92	0.33	H	0.49	0.17	0.35	0.61	0.22	0.35
Waco	674A4	TX	0.09	0.04	L	0.05	0.02	0.46	0.06	0.03	0.46
Walla Walla	687	WA	0.46	0.13	MH	0.25	0.07	0.28	0.31	0.09	0.28
Washington	688	DC	0.15	0.05	L	0.08	0.03	0.33	0.10	0.03	0.33
West Haven	689	CT	0.24	0.06	L	0.13	0.03	0.25	0.16	0.04	0.25
West Los Angeles	691	CA	1.87	0.64	VH	1.00	0.34	0.34	1.24	0.42	0.34
West Palm Beach	548	FL	0.06	0.03	L	0.03	0.01	0.42	0.04	0.02	0.42
West Roxbury	523A4	MA	0.27	0.07	ML	0.14	0.04	0.25	0.18	0.04	0.25
West Virginia	912	WV	0.14	0.05	L	0.07	0.03	0.39	0.09	0.04	0.39

(continued)

Table A.1 (continued)

Site	Center number	State	S$_s$	S$_1$	Seismicity	Site Class A			Site Class B		
						S$_{ds}$	S$_{d1}$	T$_c$	S$_{ds}$	S$_{d1}$	T$_c$
White City	692	OR	0.59	0.26	MH	0.31	0.14	0.45	0.39	0.18	0.45
White River Junction	405	VT	0.30	0.08	ML	0.16	0.04	0.27	0.20	0.05	0.27
Wichita	589A7	KS	0.14	0.05	L	0.07	0.03	0.38	0.09	0.03	0.38
Wilkes-Barre	693	PA	0.20	0.06	L	0.11	0.03	0.29	0.13	0.04	0.29
Willamette	907	OR	0.99	0.35	H	0.53	0.19	0.35	0.66	0.23	0.35
Wilmington	460	DE	0.26	0.06	ML	0.14	0.03	0.22	0.17	0.04	0.22
Wilmington	881	NC	0.30	0.10	ML	0.16	0.05	0.33	0.20	0.07	0.33
Winchester	882	VA	0.17	0.05	L	0.09	0.03	0.32	0.11	0.04	0.32
Wood	823	WI	0.11	0.04	L	0.06	0.02	0.41	0.07	0.03	0.41
Woodlawn	824	NY	0.15	0.05	L	0.08	0.03	0.35	0.10	0.04	0.35
Zachary Taylor	883	KY	0.25	0.10	ML	0.13	0.05	0.42	0.16	0.07	0.42

Site	Center number	State	S$_s$	S$_1$	Seismicity	Site Class C			Site Class D			Site Class E		
						S$_{ds}$	S$_{d1}$	T$_c$	S$_{ds}$	S$_{d1}$	T$_c$	S$_{ds}$	S$_{d1}$	T$_c$
Abraham Lincoln	915	IL	0.19	0.07	L	0.15	0.08	0.51	0.20	0.11	0.54	0.31	0.16	0.51
Albany	528A8	NY	0.23	0.07	L	0.18	0.08	0.43	0.25	0.11	0.45	0.38	0.16	0.42
Albuquerque	501	NM	0.56	0.17	MH	0.44	0.18	0.42	0.51	0.24	0.47	0.59	0.37	0.63
Alexandria	502	LA	0.13	0.06	L	0.10	0.07	0.67	0.14	0.10	0.71	0.21	0.14	0.66
Alexandria	825	LA	0.13	0.06	L	0.10	0.07	0.68	0.14	0.10	0.71	0.21	0.14	0.67
Alexandria	826	VA	0.15	0.05	L	0.12	0.06	0.46	0.16	0.08	0.49	0.26	0.12	0.46
Allen Park	553A	MI	0.13	0.05	L	0.10	0.05	0.51	0.13	0.07	0.54	0.21	0.11	0.50
Alton	800	IL	0.48	0.15	MH	0.39	0.16	0.42	0.46	0.22	0.48	0.57	0.33	0.59
Altoona	503	PA	0.14	0.05	L	0.12	0.06	0.48	0.15	0.08	0.51	0.24	0.11	0.48

(continued)

Table A.1 (continued)

Site	Center number	State	S_s	S_1	Seismicity	Site Class C			Site Class D			Site Class E		
						S_{ds}	S_{d1}	T_c	S_{ds}	S_{d1}	T_c	S_{ds}	S_{d1}	T_c
Amarillo	504	TX	0.18	0.04	L	0.14	0.05	0.35	0.19	0.07	0.37	0.29	0.10	0.34
American Lake	663A4	WA	1.19	0.42	H	0.80	0.38	0.48	0.81	0.44	0.54	0.72	0.67	0.93
Anchorage	463	AK	1.50	0.56	VH	1.00	0.49	0.49	1.00	0.56	0.56	0.90	0.90	1.00
Ann Arbor	506	MI	0.12	0.05	L	0.10	0.05	0.53	0.13	0.07	0.56	0.20	0.11	0.53
Annapolis	801	MD	0.16	0.05	L	0.12	0.06	0.46	0.17	0.08	0.48	0.26	0.12	0.45
Asheville	637	NC	0.39	0.11	MH	0.31	0.12	0.39	0.39	0.17	0.44	0.53	0.25	0.46
Aspinwall	646A4	PA	0.13	0.05	L	0.10	0.05	0.54	0.13	0.08	0.58	0.21	0.11	0.54
Atlanta	508	GA	0.23	0.09	L	0.18	0.10	0.53	0.25	0.14	0.56	0.38	0.20	0.52
Augusta	509	GA	0.37	0.11	MH	0.30	0.13	0.43	0.37	0.18	0.48	0.52	0.26	0.50
Augusta (Lenwood)	509A0	GA	0.38	0.12	MH	0.30	0.13	0.43	0.38	0.18	0.48	0.53	0.26	0.50
Balls Bluff	827	VA	0.16	0.05	L	0.13	0.06	0.45	0.17	0.08	0.48	0.27	0.12	0.45
Baltimore	512	MD	0.17	0.05	L	0.14	0.06	0.43	0.18	0.08	0.45	0.28	0.12	0.42
Baltimore	802	MD	0.17	0.05	L	0.14	0.06	0.43	0.18	0.08	0.45	0.28	0.12	0.42
Baltimore/Loch Raven	512GD	MD	0.17	0.05	L	0.14	0.06	0.43	0.18	0.08	0.45	0.28	0.12	0.42
Barrancas	828	FL	0.10	0.05	L	0.08	0.05	0.69	0.11	0.08	0.73	0.17	0.11	0.68
Batavia	528A4	NY	0.27	0.06	ML	0.21	0.07	0.32	0.28	0.10	0.34	0.43	0.14	0.32
Bath	803	NY	0.17	0.05	L	0.13	0.06	0.46	0.18	0.08	0.48	0.28	0.12	0.45
Bath	528A6	NY	0.17	0.05	L	0.13	0.06	0.46	0.18	0.08	0.48	0.28	0.12	0.45
Baton Rouge	829	LA	0.12	0.05	L	0.10	0.06	0.62	0.13	0.08	0.66	0.20	0.12	0.61
Battle Creek	515	MI	0.11	0.05	L	0.09	0.05	0.59	0.12	0.08	0.63	0.19	0.11	0.59
Bay Pines	516	FL	0.08	0.03	L	0.06	0.04	0.58	0.08	0.05	0.62	0.13	0.07	0.57

(continued)

Table A.1 (continued)

Site	Center number	State	S_s	S_1	Seismicity	Site Class C			Site Class D			Site Class E		
						S_{ds}	S_{d1}	T_c	S_{ds}	S_{d1}	T_c	S_{ds}	S_{d1}	T_c
Bay Pines	830	FL	0.08	0.03	L	0.06	0.04	0.58	0.08	0.05	0.62	0.13	0.07	0.57
Beaufort	831	SC	0.69	0.18	MH	0.52	0.20	0.38	0.58	0.25	0.44	0.61	0.40	0.65
Beckley	517	WV	0.26	0.08	ML	0.21	0.09	0.41	0.28	0.12	0.44	0.43	0.18	0.41
Bedford	518	MA	0.29	0.07	ML	0.23	0.08	0.34	0.30	0.11	0.37	0.46	0.16	0.36
Beverly	804	NJ	0.27	0.06	ML	0.22	0.07	0.31	0.29	0.10	0.33	0.44	0.14	0.32
Big Spring	519	TX	0.11	0.03	L	0.09	0.04	0.40	0.12	0.05	0.42	0.19	0.07	0.39
Biloxi	520	MS	0.12	0.05	L	0.10	0.06	0.62	0.13	0.08	0.66	0.20	0.12	0.61
Biloxi	832	MS	0.12	0.05	L	0.09	0.06	0.63	0.12	0.08	0.67	0.20	0.12	0.62
Birmingham	521	AL	0.30	0.10	ML	0.24	0.11	0.45	0.31	0.15	0.49	0.47	0.22	0.48
Black Hills	884	SD	0.15	0.04	L	0.12	0.05	0.39	0.16	0.07	0.42	0.25	0.10	0.39
Boise	531	ID	0.31	0.11	ML	0.25	0.12	0.48	0.32	0.17	0.52	0.48	0.25	0.51
Bonham	549A4	TX	0.16	0.06	L	0.13	0.07	0.55	0.17	0.10	0.58	0.27	0.14	0.54
Boston	523	MA	0.27	0.07	ML	0.22	0.08	0.35	0.29	0.11	0.37	0.44	0.16	0.35
Brevard	673GA	FL	0.08	0.04	L	0.07	0.04	0.60	0.09	0.06	0.63	0.14	0.08	0.59
Brockton	523A5	MA	0.25	0.06	ML	0.20	0.07	0.36	0.27	0.10	0.38	0.42	0.15	0.35
Bronx	526	NY	0.36	0.07	MH	0.29	0.08	0.27	0.36	0.11	0.31	0.52	0.16	0.32
Brooklyn	630A4	NY	0.35	0.07	MH	0.28	0.08	0.28	0.36	0.11	0.31	0.51	0.16	0.32
Buffalo	528	NY	0.28	0.06	ML	0.23	0.07	0.30	0.30	0.09	0.32	0.45	0.14	0.31
Butler	529	PA	0.13	0.05	L	0.10	0.05	0.53	0.14	0.08	0.56	0.21	0.11	0.53
Calverton	805	NY	0.21	0.06	L	0.17	0.06	0.38	0.22	0.09	0.40	0.35	0.13	0.38
Camp Butler	806	IL	0.27	0.11	L	0.21	0.12	0.55	0.28	0.17	0.59	0.44	0.24	0.56
Camp Nelson	833	KY	0.23	0.09	ML	0.18	0.10	0.56	0.24	0.14	0.59	0.38	0.21	0.55

(continued)

Table A.1 (continued)

Site	Center number	State	S_s	S_1	Seismicity	Site Class C			Site Class D			Site Class E		
						S_{ds}	S_{d1}	T_c	S_{ds}	S_{d1}	T_c	S_{ds}	S_{d1}	T_c
Canandaigua	528A5	NY	0.19	0.06	L	0.15	0.06	0.43	0.20	0.09	0.46	0.31	0.13	0.43
Castle Point	620A4	NY	0.28	0.07	ML	0.22	0.08	0.34	0.29	0.11	0.36	0.45	0.16	0.35
Cave Hill	834	KY	0.25	0.10	ML	0.20	0.12	0.59	0.26	0.16	0.62	0.41	0.24	0.58
Charleston	534	SC	1.44	0.36	VH	0.96	0.34	0.36	0.96	0.40	0.42	0.87	0.61	0.71
Chattanooga	835	TN	0.47	0.12	MH	0.38	0.13	0.34	0.45	0.18	0.40	0.56	0.27	0.47
Cheyenne	442	WY	0.19	0.05	L	0.15	0.06	0.39	0.20	0.08	0.41	0.32	0.12	0.39
Chicago (Lakeside)	537GD	IL	0.16	0.06	L	0.13	0.07	0.52	0.17	0.09	0.55	0.27	0.14	0.51
Chicago (Westside)	537	IL	0.17	0.06	L	0.13	0.07	0.51	0.18	0.10	0.54	0.28	0.14	0.51
Chillicothe	538	OH	0.16	0.06	L	0.13	0.07	0.58	0.17	0.10	0.61	0.26	0.15	0.57
Cincinnati	539	OH	0.18	0.08	L	0.14	0.09	0.60	0.19	0.12	0.64	0.29	0.18	0.60
City Point	836	VA	0.19	0.06	L	0.15	0.07	0.44	0.20	0.09	0.47	0.31	0.14	0.43
Clarksburg	540	WV	0.19	0.07	L	0.15	0.08	0.52	0.20	0.11	0.55	0.31	0.16	0.51
Cleveland/Brecksville	541A0	OH	0.20	0.05	L	0.16	0.06	0.37	0.21	0.08	0.40	0.33	0.12	0.37
Cleveland/Wade Park	541	OH	0.20	0.05	L	0.16	0.06	0.37	0.21	0.08	0.40	0.33	0.12	0.37
Coatesville	542	PA	0.27	0.06	ML	0.22	0.07	0.31	0.29	0.10	0.33	0.44	0.14	0.32
Cold Harbor	837	VA	0.21	0.06	L	0.16	0.07	0.40	0.22	0.09	0.42	0.34	0.14	0.39
Columbia	589A4	MO	0.20	0.09	L	0.16	0.10	0.63	0.21	0.14	0.67	0.33	0.21	0.62
Columbia	544	SC	0.57	0.15	MH	0.45	0.17	0.38	0.51	0.22	0.44	0.59	0.34	0.57
Corinth	838	MS	0.50	0.17	MH	0.40	0.18	0.46	0.47	0.24	0.51	0.57	0.37	0.65
Crown Hill	807	IN	0.19	0.08	L	0.15	0.09	0.62	0.20	0.13	0.66	0.32	0.19	0.61
Culpeper	839	VA	0.19	0.06	L	0.15	0.06	0.42	0.21	0.09	0.44	0.32	0.13	0.41
Cypress Hills	808	NY	0.36	0.07	MH	0.29	0.08	0.27	0.36	0.11	0.31	0.51	0.16	0.31

(continued)

Table A.1 (continued)

Site	Center number	State	S_s	S_1	Seismicity	Site Class C			Site Class D			Site Class E		
						S_{ds}	S_{d1}	T_c	S_{ds}	S_{d1}	T_c	S_{ds}	S_{d1}	T_c
Dallas	549	TX	0.11	0.05	L	0.09	0.06	0.61	0.12	0.08	0.65	0.19	0.11	0.61
Dallas/Fort Worth	916	TX	0.12	0.05	L	0.09	0.06	0.61	0.12	0.08	0.65	0.19	0.12	0.60
Danville	550	IL	0.22	0.09	L	0.18	0.10	0.58	0.24	0.15	0.61	0.37	0.21	0.57
Danville	809	IL	0.22	0.09	L	0.18	0.10	0.58	0.24	0.15	0.61	0.37	0.21	0.57
Danville	840	KY	0.22	0.09	L	0.18	0.10	0.60	0.23	0.15	0.63	0.37	0.21	0.59
Danville	841	VA	0.20	0.07	L	0.16	0.08	0.53	0.21	0.12	0.57	0.33	0.17	0.53
Dayton	552	OH	0.19	0.07	L	0.15	0.08	0.53	0.20	0.11	0.56	0.31	0.16	0.52
Dayton	810	OH	0.21	0.07	L	0.16	0.08	0.48	0.22	0.11	0.51	0.34	0.16	0.48
Denver	554	CO	0.21	0.06	L	0.17	0.06	0.37	0.23	0.09	0.39	0.36	0.13	0.37
Des Moines	636A6	IA	0.08	0.04	L	0.06	0.05	0.81	0.08	0.07	0.86	0.13	0.10	0.80
Detroit	553	MI	0.12	0.05	L	0.10	0.05	0.53	0.13	0.07	0.56	0.20	0.11	0.52
Dublin	557	GA	0.22	0.08	L	0.17	0.09	0.55	0.23	0.13	0.58	0.36	0.19	0.54
Durham	558	NC	0.20	0.08	L	0.16	0.09	0.56	0.21	0.12	0.59	0.33	0.18	0.55
Eagle Point	906	OR	0.58	0.26	MH	0.45	0.26	0.58	0.52	0.32	0.62	0.60	0.51	0.85
East Orange	561	NJ	0.36	0.07	MH	0.29	0.08	0.28	0.37	0.11	0.31	0.52	0.17	0.32
El Paso	756	TX	0.33	0.11	ML	0.27	0.12	0.45	0.34	0.17	0.50	0.50	0.25	0.50
Erie	562	PA	0.16	0.05	L	0.13	0.06	0.43	0.17	0.08	0.45	0.27	0.11	0.42
Fargo	437	ND	0.07	0.02	L	0.06	0.02	0.40	0.08	0.03	0.43	0.12	0.05	0.40
Fayetteville	564	AR	0.21	0.09	L	0.17	0.10	0.62	0.22	0.15	0.66	0.35	0.21	0.62
Fayetteville	565	NC	0.30	0.10	ML	0.24	0.11	0.48	0.31	0.16	0.52	0.47	0.24	0.51
Fayetteville	842	AR	0.21	0.09	L	0.17	0.10	0.62	0.22	0.15	0.66	0.35	0.21	0.62
Finn's Point	811	NJ	0.23	0.06	L	0.19	0.06	0.34	0.25	0.09	0.36	0.39	0.13	0.33

(continued)

Table A.1 (continued)

Site	Center number	State	S_s	S_1	Seismicity	Site Class C			Site Class D			Site Class E		
						S_{ds}	S_{d1}	T_c	S_{ds}	S_{d1}	T_c	S_{ds}	S_{d1}	T_c
Florence	843	SC	0.73	0.20	MH	0.54	0.21	0.39	0.59	0.27	0.45	0.60	0.42	0.70
Florida	911	FL	0.09	0.04	L	0.07	0.04	0.60	0.10	0.06	0.63	0.15	0.09	0.59
Fort Bayard	885	NM	0.27	0.08	ML	0.22	0.09	0.42	0.29	0.13	0.45	0.44	0.19	0.43
Fort Bliss	886	TX	0.34	0.11	MH	0.27	0.12	0.46	0.34	0.17	0.50	0.50	0.25	0.51
Fort Custer	909	MI	0.11	0.05	L	0.09	0.05	0.61	0.12	0.08	0.65	0.18	0.11	0.60
Fort Gibson	844	OK	0.19	0.08	L	0.15	0.09	0.57	0.20	0.12	0.60	0.31	0.18	0.56
Fort Harrison	436	MT	0.75	0.22	MH	0.55	0.23	0.43	0.60	0.29	0.48	0.60	0.46	0.77
Fort Harrison	845	VA	0.23	0.06	L	0.18	0.07	0.38	0.24	0.10	0.40	0.38	0.14	0.37
Fort Howard	512GF	MD	0.17	0.05	L	0.13	0.06	0.42	0.18	0.08	0.45	0.28	0.12	0.42
Fort Leavenworth	887	KS	0.13	0.06	L	0.10	0.06	0.60	0.14	0.09	0.64	0.22	0.13	0.60
Fort Logan	888	CO	0.22	0.06	L	0.18	0.06	0.37	0.23	0.09	0.39	0.37	0.13	0.36
Fort Lyon	567	CO	0.17	0.05	L	0.13	0.06	0.43	0.18	0.08	0.45	0.28	0.12	0.42
Fort Lyon	889	CO	0.17	0.05	L	0.13	0.06	0.43	0.18	0.08	0.45	0.28	0.12	0.42
Fort McPherson	890	NE	0.09	0.03	L	0.08	0.04	0.50	0.10	0.05	0.53	0.16	0.08	0.49
Fort Meade	568	SD	0.21	0.05	L	0.16	0.06	0.35	0.22	0.08	0.37	0.34	0.12	0.35
Fort Meade	891	SD	0.21	0.05	L	0.16	0.06	0.35	0.22	0.08	0.37	0.34	0.12	0.35
Fort Mitchell	908	AL	0.14	0.07	L	0.11	0.07	0.66	0.15	0.11	0.70	0.24	0.15	0.65
Fort Richardson	910	AK	1.50	0.56	VH	1.00	0.49	0.48	1.00	0.56	0.56	0.90	0.90	0.99
Fort Rosecrans	892	CA	1.57	0.61	VH	1.05	0.53	0.51	1.05	0.61	0.59	0.94	0.98	1.04
Fort Sam Houston	846	TX	0.11	0.03	L	0.08	0.03	0.39	0.11	0.05	0.41	0.18	0.07	0.39
Fort Scott	893	KS	0.13	0.07	L	0.10	0.07	0.73	0.14	0.11	0.77	0.21	0.15	0.72
Fort Sill	920	OK	0.37	0.09	MH	0.30	0.10	0.32	0.37	0.14	0.36	0.52	0.20	0.38

(continued)

Table A.1 (continued)

Site	Center number	State	S_s	S_1	Seismicity	Site Class C			Site Class D			Site Class E		
						S_{ds}	S_{d1}	T_c	S_{ds}	S_{d1}	T_c	S_{ds}	S_{d1}	T_c
Fort Smith	847	AR	0.21	0.09	L	0.17	0.10	0.60	0.22	0.14	0.63	0.35	0.21	0.59
Fort Snelling	894	MN	0.06	0.03	L	0.05	0.03	0.63	0.07	0.04	0.66	0.10	0.06	0.62
Fort Thomas	539A	OH	0.15	0.06	L	0.12	0.07	0.55	0.16	0.09	0.58	0.25	0.14	0.54
Fort Wayne	610A4	IN	0.15	0.06	L	0.12	0.07	0.56	0.16	0.09	0.59	0.25	0.14	0.55
Fresno	570	CA	0.50	0.22	MH	0.40	0.23	0.58	0.47	0.29	0.62	0.57	0.46	0.81
Gainesville	573	FL	0.11	0.05	L	0.09	0.05	0.64	0.11	0.08	0.67	0.18	0.11	0.63
Glendale	848	VA	0.23	0.06	L	0.18	0.07	0.38	0.24	0.10	0.40	0.38	0.14	0.37
Golden Gate	895	CA	2.22	1.27	VH	1.48	1.10	0.74	1.48	1.27	0.86	1.33	2.03	1.52
Grafton	812	WV	0.14	0.05	L	0.11	0.06	0.55	0.15	0.09	0.58	0.23	0.13	0.54
Grand Island	636A4	NE	0.13	0.04	L	0.10	0.04	0.43	0.14	0.06	0.45	0.22	0.09	0.42
Grand Junction	575	CO	0.29	0.07	ML	0.23	0.08	0.33	0.30	0.11	0.36	0.46	0.16	0.34
Gulfport	520A0	MS	0.12	0.05	L	0.10	0.06	0.62	0.13	0.08	0.66	0.20	0.12	0.61
Hampton	590	VA	0.12	0.05	L	0.10	0.05	0.57	0.13	0.08	0.60	0.20	0.11	0.56
Hampton	849	VA	0.12	0.05	L	0.10	0.06	0.57	0.13	0.08	0.60	0.20	0.11	0.56
Hampton (VAMC)	850	VA	0.12	0.05	L	0.10	0.06	0.57	0.13	0.08	0.60	0.20	0.11	0.56
Hines	578	IL	0.17	0.06	L	0.14	0.07	0.47	0.18	0.09	0.50	0.29	0.14	0.47
Hines VBA	201	IL	0.17	0.06	L	0.14	0.07	0.47	0.18	0.09	0.50	0.29	0.14	0.47
Honolulu**	459	HI	0.61	0.18	MH	0.47	0.19	0.41	0.54	0.25	0.46	0.60	0.39	0.64
Hot Springs	896	SD	0.21	0.05	L	0.17	0.06	0.34	0.22	0.08	0.36	0.35	0.11	0.33
Hot Springs	568A4	SD	0.21	0.05	L	0.17	0.06	0.34	0.22	0.08	0.36	0.35	0.11	0.33
Houston	580	TX	0.09	0.04	L	0.07	0.04	0.59	0.09	0.06	0.62	0.15	0.08	0.58
Houston	851	TX	0.09	0.04	L	0.07	0.04	0.58	0.09	0.06	0.61	0.15	0.08	0.57

(continued)

Table A.1 (continued)

Site	Center number	State	S_s	S_1	Seismicity	Site Class C			Site Class D			Site Class E		
						S_{ds}	S_{d1}	T_c	S_{ds}	S_{d1}	T_c	S_{ds}	S_{d1}	T_c
Houston VBA	362	TX	0.09	0.04	L	0.07	0.04	0.59	0.09	0.06	0.62	0.15	0.08	0.58
Huntington	581	WV	0.19	0.07	L	0.15	0.08	0.52	0.20	0.11	0.55	0.32	0.17	0.52
Indianapolis	583	IN	0.19	0.08	L	0.15	0.09	0.61	0.20	0.13	0.65	0.32	0.19	0.61
Indianapolis (CS Rd)	583A4	IN	0.19	0.08	L	0.15	0.09	0.61	0.20	0.13	0.65	0.32	0.19	0.61
Indiantown Gap	813	PA	0.22	0.06	L	0.18	0.06	0.36	0.23	0.09	0.38	0.37	0.13	0.36
Iowa City	636A8	IA	0.10	0.05	L	0.08	0.06	0.74	0.11	0.08	0.78	0.17	0.12	0.73
Iron Mountain	585	MI	0.06	0.03	L	0.05	0.03	0.64	0.06	0.04	0.67	0.10	0.06	0.63
Jackson	586	MS	0.19	0.09	L	0.16	0.10	0.63	0.21	0.14	0.66	0.32	0.20	0.62
Jackson VBA	323	MS	0.19	0.09	L	0.16	0.10	0.63	0.21	0.14	0.66	0.32	0.20	0.62
Jefferson Barracks	852	MO	0.58	0.17	MH	0.45	0.18	0.40	0.52	0.24	0.46	0.60	0.37	0.62
Jefferson City	853	MO	0.24	0.10	ML	0.19	0.11	0.60	0.25	0.16	0.64	0.39	0.23	0.59
Kansas City	589	MO	0.13	0.06	L	0.10	0.07	0.66	0.14	0.09	0.70	0.21	0.14	0.65
Keokuk	814	IA	0.15	0.07	L	0.12	0.08	0.69	0.16	0.12	0.73	0.25	0.17	0.68
Kerrville	854	TX	0.07	0.03	L	0.06	0.03	0.50	0.08	0.04	0.53	0.12	0.06	0.49
Kerrville	671A4	TX	0.07	0.03	L	0.06	0.03	0.50	0.08	0.04	0.53	0.12	0.06	0.49
Knoxville	636A7	IA	0.08	0.05	L	0.07	0.05	0.82	0.09	0.08	0.87	0.14	0.11	0.81
Knoxville	855	TN	0.52	0.12	MH	0.41	0.13	0.32	0.48	0.18	0.38	0.58	0.27	0.47
Lake City	573A4	FL	0.12	0.05	L	0.10	0.06	0.63	0.13	0.09	0.66	0.20	0.13	0.62
Las Vegas	593	NV	0.55	0.17	MH	0.43	0.19	0.43	0.50	0.24	0.48	0.59	0.37	0.64
Leavenworth	897	KS	0.13	0.06	L	0.10	0.06	0.60	0.14	0.09	0.64	0.22	0.13	0.60
Leavenworth	589A6	KS	0.13	0.06	L	0.10	0.06	0.60	0.14	0.09	0.64	0.22	0.13	0.60
Lebanon	856	KY	0.23	0.10	L	0.18	0.11	0.62	0.24	0.16	0.65	0.38	0.23	0.61

(continued)

Table A.1 (continued)

Site	Center number	State	S_s	S_1	Seismicity	Site Class C			Site Class D			Site Class E		
						S_{ds}	S_{d1}	T_c	S_{ds}	S_{d1}	T_c	S_{ds}	S_{d1}	T_c
Lebanon	595	PA	0.23	0.06	L	0.18	0.06	0.35	0.24	0.09	0.38	0.38	0.13	0.35
Lexington	857	KY	0.22	0.09	L	0.18	0.10	0.56	0.24	0.14	0.59	0.37	0.21	0.55
Lexington (CD)	596A4	KY	0.23	0.09	L	0.18	0.10	0.54	0.24	0.14	0.57	0.38	0.20	0.53
Lexington (LD)	596	KY	0.23	0.09	L	0.18	0.10	0.54	0.24	0.14	0.57	0.38	0.20	0.53
Lincoln	636A5	NE	0.18	0.05	L	0.14	0.05	0.37	0.19	0.07	0.39	0.30	0.11	0.36
Little Rock	598	AR	0.49	0.16	MH	0.40	0.17	0.44	0.46	0.23	0.50	0.57	0.35	0.63
Little Rock	858	AR	0.51	0.16	MH	0.40	0.18	0.44	0.47	0.23	0.50	0.57	0.36	0.63
Livermore	640A4	CA	1.59	0.60	VH	1.06	0.52	0.49	1.06	0.60	0.57	0.95	0.96	1.01
Loma Linda	605	CA	1.76	0.61	VH	1.17	0.53	0.45	1.17	0.61	0.52	1.06	0.98	0.92
Long Beach	600	CA	2.02	0.85	VH	1.35	0.74	0.55	1.35	0.85	0.63	1.21	1.36	1.12
Long Island	815	NY	0.29	0.06	ML	0.23	0.07	0.30	0.31	0.10	0.33	0.46	0.15	0.32
Los Angeles	898	CA	1.66	0.59	VH	1.10	0.51	0.46	1.10	0.59	0.53	0.99	0.94	0.95
Los Angeles	691GE	CA	2.23	0.77	L	1.49	0.66	0.45	1.49	0.77	0.51	1.34	1.23	0.92
Loudon Park	816	MD	0.17	0.05	ML	0.14	0.06	0.43	0.18	0.08	0.45	0.28	0.12	0.42
Louisville	603	KY	0.25	0.10	ML	0.20	0.12	0.59	0.26	0.16	0.62	0.41	0.24	0.58
Lyons	561A4	NJ	0.35	0.07	ML	0.28	0.08	0.28	0.35	0.11	0.31	0.51	0.16	0.32
Madison	607	WI	0.10	0.04	L	0.08	0.05	0.60	0.11	0.07	0.63	0.17	0.10	0.59
Manchester	608	NH	0.35	0.08	MH	0.28	0.09	0.32	0.36	0.13	0.36	0.51	0.19	0.37
Marietta	859	GA	0.25	0.09	ML	0.20	0.10	0.50	0.27	0.14	0.53	0.42	0.21	0.50
Marion	657A5	IL	1.12	0.31	H	0.75	0.30	0.41	0.78	0.36	0.46	0.67	0.57	0.84
Marion	610	IN	0.15	0.07	L	0.12	0.08	0.63	0.16	0.11	0.67	0.25	0.16	0.62
Marion	817	IN	0.15	0.07	L	0.12	0.08	0.63	0.16	0.11	0.67	0.25	0.16	0.62

(continued)

Table A.1 (continued)

Site	Center number	State	S$_s$	S$_1$	Seismicity	Site Class C			Site Class D			Site Class E		
						S$_{ds}$	S$_{d1}$	T$_c$	S$_{ds}$	S$_{d1}$	T$_c$	S$_{ds}$	S$_{d1}$	T$_c$
Marlin	674A5	TX	0.09	0.04	L	0.07	0.05	0.63	0.10	0.06	0.67	0.15	0.09	0.62
Martinez/NCSC	612	CA	1.58	0.60	VH	1.05	0.52	0.50	1.05	0.60	0.57	0.95	0.96	1.02
Martinsburg	613	WV	0.17	0.05	L	0.13	0.06	0.44	0.18	0.08	0.47	0.28	0.12	0.44
Massachusetts	818	MA	0.21	0.06	L	0.17	0.06	0.38	0.23	0.09	0.40	0.35	0.13	0.37
McClellan	612GH	CA	0.49	0.22	MH	0.39	0.23	0.60	0.46	0.29	0.63	0.57	0.46	0.81
Memphis	614	TN	1.29	0.35	VH	0.86	0.34	0.40	0.86	0.40	0.46	0.77	0.61	0.79
Memphis	860	TN	1.29	0.35	VH	0.86	0.34	0.40	0.86	0.40	0.46	0.77	0.61	0.79
Menlo Park	640A0	CA	1.79	0.79	VH	1.19	0.69	0.58	1.19	0.79	0.66	1.08	1.27	1.18
Miami	546	FL	0.05	0.02	L	0.04	0.02	0.53	0.05	0.03	0.56	0.09	0.04	0.52
Miles City	436GJ	MT	0.10	0.03	L	0.08	0.04	0.49	0.11	0.05	0.52	0.17	0.08	0.48
Mill Springs	861	KY	0.23	0.10	L	0.19	0.11	0.58	0.25	0.15	0.62	0.39	0.22	0.57
Milwaukee (Wood)	695	WI	0.11	0.05	L	0.09	0.05	0.59	0.12	0.07	0.63	0.18	0.11	0.58
Minneapolis	618	MN	0.06	0.03	L	0.05	0.03	0.64	0.06	0.04	0.68	0.10	0.06	0.63
Mobile	862	AL	0.12	0.05	L	0.09	0.06	0.64	0.12	0.08	0.68	0.20	0.12	0.63
Montgomery	619	AL	0.15	0.07	L	0.12	0.08	0.63	0.16	0.11	0.67	0.26	0.16	0.63
Montgomery VBA	322	AL	0.15	0.07	L	0.12	0.08	0.63	0.16	0.11	0.67	0.26	0.16	0.63
Montrose	620	NY	0.33	0.07	ML	0.27	0.08	0.30	0.34	0.11	0.33	0.50	0.16	0.33
Mound City	863	IL	3.39	1.31	VH	2.26	1.14	0.50	2.26	1.31	0.58	2.03	2.10	1.03
Mountain Home	621	TN	0.39	0.10	MH	0.31	0.12	0.37	0.39	0.16	0.42	0.53	0.24	0.44
Mountain Home	864	TN	0.39	0.10	MH	0.31	0.12	0.37	0.39	0.16	0.42	0.53	0.24	0.44
Murfreesboro	626A4	TN	0.29	0.12	ML	0.23	0.13	0.56	0.30	0.18	0.60	0.46	0.27	0.58
Muskogee	623	OK	0.19	0.07	L	0.15	0.08	0.57	0.20	0.12	0.60	0.31	0.17	0.56

(continued)

Table A.1 (continued)

Site	Center number	State	S_s	S_1	Seismicity	Site Class C			Site Class D			Site Class E		
						S_{ds}	S_{d1}	T_c	S_{ds}	S_{d1}	T_c	S_{ds}	S_{d1}	T_c
Nashville	626	TN	0.35	0.13	ML	0.28	0.15	0.54	0.35	0.20	0.57	0.51	0.30	0.60
Nashville	865	TN	0.35	0.13	ML	0.28	0.15	0.54	0.35	0.20	0.57	0.51	0.30	0.60
Natchez	866	MS	0.14	0.07	L	0.11	0.08	0.67	0.15	0.11	0.71	0.24	0.16	0.67
NCA Operations Support	786	VA	0.19	0.07	L	0.15	0.07	0.49	0.20	0.11	0.52	0.32	0.15	0.49
New Albany	867	IN	0.25	0.10	ML	0.20	0.12	0.58	0.27	0.16	0.61	0.42	0.24	0.58
New Bern	868	NC	0.16	0.07	L	0.13	0.07	0.57	0.17	0.10	0.60	0.27	0.15	0.56
New Orleans	629	LA	0.11	0.05	L	0.09	0.05	0.62	0.12	0.08	0.65	0.18	0.11	0.61
New York	630	NY	0.36	0.07	MH	0.29	0.08	0.27	0.36	0.11	0.31	0.52	0.16	0.32
Newington	689A4	CT	0.24	0.06	L	0.19	0.07	0.37	0.26	0.10	0.39	0.40	0.15	0.37
NMCA	914	AZ	0.18	0.06	L	0.14	0.07	0.49	0.19	0.10	0.51	0.30	0.14	0.48
NMCP	899	HI	0.61	0.18	MH	0.47	0.19	0.41	0.54	0.25	0.46	0.60	0.39	0.64
North Chicago	556	IL	0.14	0.05	L	0.11	0.06	0.53	0.15	0.08	0.56	0.24	0.12	0.53
North Little Rock	598A0	AR	0.51	0.17	MH	0.41	0.18	0.44	0.48	0.24	0.49	0.57	0.36	0.63
Northampton	631	MA	0.22	0.07	LL	0.18	0.07	0.42	0.24	0.11	0.44	0.37	0.15	0.41
Northport	632	NY	0.29	0.06	ML	0.23	0.07	0.32	0.30	0.10	0.34	0.45	0.15	0.33
Oklahoma City	635	OK	0.34	0.07	ML	0.27	0.08	0.31	0.34	0.12	0.35	0.50	0.17	0.35
Omaha	636	NE	0.12	0.04	L	0.10	0.05	0.48	0.13	0.07	0.51	0.21	0.10	0.48
Orlando	673BY	FL	0.10	0.04	L	0.08	0.04	0.56	0.10	0.06	0.59	0.16	0.09	0.55
Palo Alto	640	CA	1.96	0.83	VH	1.31	0.72	0.55	1.31	0.83	0.63	1.18	1.32	1.13
Perry Point	512A5	MD	0.22	0.05	L	0.17	0.06	0.35	0.23	0.09	0.38	0.36	0.13	0.35
Philadelphia	642	PA	0.27	0.06	ML	0.22	0.07	0.31	0.29	0.10	0.34	0.44	0.14	0.32

(continued)

Table A.1 (continued)

Site	Center number	State	S_s	S_1	Seismicity	Site Class C			Site Class D			Site Class E		
						S_{ds}	S_{d1}	T_c	S_{ds}	S_{d1}	T_c	S_{ds}	S_{d1}	T_c
Philadelphia	819	PA	0.28	0.06	ML	0.22	0.07	0.31	0.30	0.10	0.34	0.45	0.14	0.32
Phoenix	644	AZ	0.18	0.06	L	0.15	0.07	0.48	0.19	0.10	0.51	0.30	0.14	0.48
Pittsburgh (HD)	646A5	PA	0.13	0.05	L	0.10	0.06	0.56	0.13	0.08	0.59	0.21	0.11	0.55
Pittsburgh (UD)	646	PA	0.13	0.05	L	0.10	0.05	0.54	0.13	0.08	0.58	0.21	0.11	0.54
Poplar Bluff	657A4	MO	1.10	0.30	H	0.73	0.30	0.41	0.78	0.36	0.47	0.66	0.56	0.85
Port Hudson	870	LA	0.12	0.06	L	0.10	0.06	0.63	0.13	0.09	0.67	0.21	0.13	0.63
Portland	648	OR	0.98	0.35	H	0.66	0.33	0.51	0.73	0.39	0.54	0.60	0.60	1.00
Prescott	649	AZ	0.34	0.10	ML	0.27	0.11	0.41	0.35	0.16	0.46	0.50	0.23	0.46
Prescott	900	AZ	0.35	0.10	MH	0.28	0.12	0.41	0.35	0.16	0.46	0.51	0.24	0.47
Providence	650	RI	0.23	0.06	L	0.19	0.07	0.37	0.25	0.10	0.39	0.39	0.14	0.36
Quantico	872	VA	0.16	0.05	L	0.13	0.06	0.45	0.17	0.08	0.48	0.27	0.12	0.45
Quincy	820	IL	0.18	0.08	L	0.14	0.09	0.64	0.19	0.13	0.68	0.30	0.19	0.63
Raleigh	873	NC	0.20	0.08	L	0.16	0.09	0.55	0.22	0.13	0.59	0.34	0.18	0.55
Reno	654	NV	1.50	0.60	VH	1.00	0.52	0.52	1.00	0.60	0.60	0.90	0.96	1.07
Richmond	652	VA	0.23	0.06	L	0.18	0.07	0.38	0.24	0.10	0.40	0.38	0.14	0.37
Richmond	874	VA	0.23	0.06	L	0.18	0.07	0.38	0.24	0.10	0.40	0.38	0.14	0.37
Riverside	901	CA	1.50	0.60	VH	1.00	0.52	0.52	1.00	0.60	0.60	0.90	0.96	1.07
Rock Island	821	IL	0.13	0.06	L	0.10	0.07	0.65	0.14	0.10	0.69	0.22	0.14	0.64
Roseburg	653	OR	0.83	0.42	VH	0.59	0.39	0.66	0.65	0.44	0.69	0.61	0.68	1.11
Roseburg	902	OR	0.83	0.42	H	0.59	0.39	0.66	0.65	0.44	0.69	0.61	0.68	1.11
Sacramento NCHCS	612A4	CA	0.46	0.21	MH	0.37	0.23	0.61	0.44	0.28	0.64	0.56	0.45	0.80
Saginaw	655	MI	0.08	0.04	L	0.06	0.04	0.66	0.09	0.06	0.69	0.13	0.09	0.65

(continued)

Table A.1 (continued)

Site	Center number	State	S_s	S_1	Seismicity	Site Class C			Site Class D			Site Class E		
						S_{ds}	S_{d1}	T_c	S_{ds}	S_{d1}	T_c	S_{ds}	S_{d1}	T_c
Salem	658	VA	0.26	0.08	ML	0.21	0.09	0.41	0.28	0.12	0.43	0.43	0.18	0.41
Salisbury	659	NC	0.26	0.09	ML	0.21	0.11	0.51	0.28	0.15	0.54	0.43	0.22	0.51
Salisbury	876	NC	0.26	0.09	ML	0.21	0.11	0.51	0.28	0.15	0.54	0.43	0.22	0.51
Salt Lake City	660	UT	1.58	0.63	VH	1.05	0.54	0.52	1.05	0.63	0.59	0.95	1.00	1.06
San Antonio	671	TX	0.11	0.03	L	0.08	0.04	0.42	0.11	0.05	0.44	0.18	0.07	0.41
San Antonio	877	TX	0.11	0.03	L	0.08	0.03	0.39	0.11	0.05	0.41	0.18	0.07	0.39
San Diego	664	CA	1.56	0.60	VH	1.04	0.52	0.50	1.04	0.60	0.58	0.94	0.96	1.03
San Francisco	662	CA	1.76	0.90	VH	1.17	0.78	0.67	1.17	0.90	0.77	1.06	1.44	1.36
San Francisco	903	CA	1.50	0.67	VH	1.00	0.58	0.58	1.00	0.67	0.67	0.90	1.07	1.19
San Joaquin Valley	913	CA	1.83	0.60	VH	1.22	0.52	0.43	1.22	0.60	0.49	1.10	0.96	0.88
San Juan	672	PR	0.90	0.31	H	0.62	0.31	0.50	0.68	0.37	0.54	0.61	0.57	0.94
Santa Fe	904	NM	0.48	0.16	MH	0.39	0.17	0.44	0.46	0.23	0.50	0.57	0.35	0.61
Saratoga	917	NY	0.25	0.07	ML	0.20	0.08	0.42	0.27	0.12	0.44	0.42	0.17	0.41
Seattle	663	WA	1.55	0.53	VH	1.03	0.46	0.45	1.03	0.53	0.52	0.93	0.85	0.92
Sepulveda	691A4	CA	2.04	0.73	VH	1.36	0.63	0.46	1.36	0.73	0.53	1.23	1.16	0.95
Seven Pines	878	VA	0.20	0.06	L	0.16	0.06	0.41	0.21	0.09	0.43	0.33	0.13	0.40
Sheridan	666	WY	0.27	0.06	ML	0.22	0.07	0.31	0.29	0.10	0.34	0.44	0.14	0.32
Shreveport	667	LA	0.15	0.07	L	0.12	0.08	0.64	0.16	0.11	0.68	0.26	0.16	0.63
Sioux Falls	438	SD	0.11	0.03	L	0.09	0.04	0.43	0.12	0.05	0.46	0.19	0.08	0.43
Sitka	905	AK	0.97	0.50	H	0.65	0.43	0.66	0.72	0.50	0.69	0.61	0.80	1.31
Somerville AMS	796	NJ	0.33	0.07	ML	0.26	0.08	0.29	0.33	0.11	0.32	0.49	0.16	0.32
Spokane	668	WA	0.40	0.11	MH	0.32	0.13	0.40	0.40	0.18	0.45	0.54	0.26	0.49

(continued)

Table A.1 (continued)

Site	Center number	State	S_s	S_1	Seismicity	Site Class C			Site Class D			Site Class E		
						S_{ds}	S_{d1}	T_c	S_{ds}	S_{d1}	T_c	S_{ds}	S_{d1}	T_c
Springfield	879	MO	0.22	0.10	L	0.18	0.11	0.62	0.24	0.15	0.65	0.37	0.22	0.61
St. Albans	630A5	NY	0.34	0.07	ML	0.27	0.08	0.28	0.34	0.11	0.31	0.50	0.16	0.31
St. Augustine	875	FL	0.13	0.05	L	0.10	0.06	0.60	0.13	0.08	0.63	0.21	0.12	0.59
St. Cloud	656	MN	0.08	0.02	L	0.06	0.02	0.39	0.08	0.04	0.42	0.13	0.05	0.39
St. Louis (JB)	657A0	MO	0.60	0.17	MH	0.46	0.19	0.40	0.53	0.24	0.46	0.60	0.37	0.63
St. Louis (JC)	657	MO	0.60	0.17	MH	0.46	0.19	0.40	0.53	0.24	0.46	0.60	0.37	0.63
St. Petersburg VBA	317	FL	0.08	0.03	L	0.06	0.04	0.58	0.08	0.05	0.62	0.13	0.07	0.57
Staunton	880	VA	0.21	0.06	L	0.17	0.07	0.42	0.23	0.10	0.45	0.35	0.15	0.42
Syracuse	528A7	NY	0.18	0.06	L	0.14	0.07	0.48	0.19	0.10	0.51	0.30	0.14	0.47
Tahoma	919	WA	1.28	0.44	VH	0.86	0.40	0.46	0.86	0.45	0.53	0.77	0.70	0.91
Tampa	673	FL	0.08	0.03	L	0.06	0.04	0.59	0.08	0.05	0.62	0.13	0.07	0.58
Temple	674	TX	0.08	0.04	L	0.07	0.04	0.64	0.09	0.06	0.68	0.14	0.09	0.63
Togus	402	ME	0.29	0.08	ML	0.23	0.09	0.37	0.30	0.12	0.40	0.46	0.18	0.39
Togus	822	ME	0.29	0.08	ML	0.23	0.09	0.37	0.30	0.12	0.40	0.46	0.18	0.39
Tomah	676	WI	0.07	0.03	L	0.05	0.04	0.72	0.07	0.05	0.76	0.11	0.08	0.71
Topeka	589A5	KS	0.16	0.05	L	0.13	0.06	0.49	0.17	0.09	0.52	0.26	0.13	0.48
Tucson	678	AZ	0.29	0.08	ML	0.23	0.09	0.40	0.30	0.13	0.43	0.46	0.19	0.41
Tuscaloosa	679	AL	0.27	0.09	ML	0.21	0.11	0.49	0.28	0.15	0.53	0.44	0.22	0.50
Tuskegee	619A4	AL	0.15	0.07	L	0.12	0.08	0.64	0.16	0.11	0.68	0.25	0.16	0.63
Vancouver	648A4	WA	0.92	0.33	H	0.63	0.32	0.51	0.69	0.38	0.55	0.61	0.59	0.96
Waco	674A4	TX	0.09	0.04	L	0.07	0.05	0.65	0.09	0.06	0.69	0.15	0.09	0.64
Walla Walla	687	WA	0.46	0.13	MH	0.37	0.15	0.40	0.44	0.20	0.45	0.56	0.30	0.53

(continued)

Table A.1 (continued)

Site	Center number	State	S_s	S_1	Seismicity	Site Class C			Site Class D			Site Class E		
						S_{ds}	S_{d1}	T_c	S_{ds}	S_{d1}	T_c	S_{ds}	S_{d1}	T_c
Washington	688	DC	0.15	0.05	L	0.12	0.06	0.46	0.16	0.08	0.49	0.26	0.12	0.46
West Haven	689	CT	0.24	0.06	L	0.20	0.07	0.36	0.26	0.10	0.38	0.41	0.14	0.36
West Los Angeles	691	CA	1.87	0.64	VH	1.24	0.55	0.44	1.24	0.64	0.51	1.12	1.02	0.91
West Palm Beach	548	FL	0.06	0.03	L	0.05	0.03	0.59	0.06	0.04	0.63	0.10	0.06	0.58
West Roxbury	523A4	MA	0.27	0.07	ML	0.21	0.07	0.35	0.28	0.11	0.38	0.43	0.15	0.36
West Virginia	912	WV	0.14	0.05	L	0.11	0.06	0.55	0.15	0.09	0.58	0.23	0.13	0.54
White City	692	OR	0.59	0.26	MH	0.45	0.27	0.59	0.52	0.33	0.63	0.60	0.52	0.87
White River Junction	405	VT	0.30	0.08	ML	0.24	0.09	0.38	0.31	0.13	0.42	0.47	0.19	0.40
Wichita	589A7	KS	0.14	0.05	L	0.11	0.06	0.54	0.14	0.08	0.57	0.23	0.12	0.53
Wilkes-Barre	693	PA	0.20	0.06	L	0.16	0.06	0.41	0.21	0.09	0.43	0.33	0.13	0.40
Willamette	907	OR	0.99	0.35	H	0.66	0.34	0.51	0.73	0.40	0.54	0.60	0.61	1.00
Wilmington	460	DE	0.26	0.06	ML	0.21	0.07	0.32	0.28	0.09	0.34	0.43	0.14	0.32
Wilmington	881	NC	0.30	0.10	ML	0.24	0.11	0.47	0.31	0.16	0.51	0.46	0.23	0.49
Winchester	882	VA	0.17	0.05	L	0.13	0.06	0.46	0.18	0.09	0.49	0.28	0.13	0.45
Wood	823	WI	0.11	0.04	L	0.09	0.05	0.58	0.11	0.07	0.62	0.18	0.10	0.58
Woodlawn	824	NY	0.15	0.05	L	0.12	0.06	0.49	0.16	0.08	0.52	0.26	0.12	0.48
Zachary Taylor	883	KY	0.25	0.10	ML	0.20	0.12	0.59	0.26	0.16	0.62	0.41	0.24	0.58

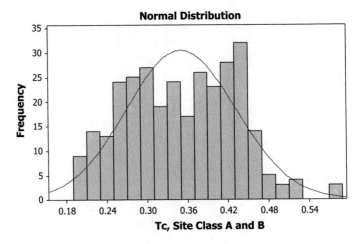

Figure A.1 Data frequency and normal distribution curves for site class A and B

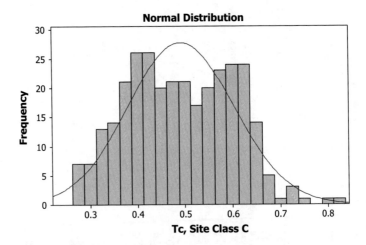

Figure A.2 Data frequency and normal distribution curves for site class C

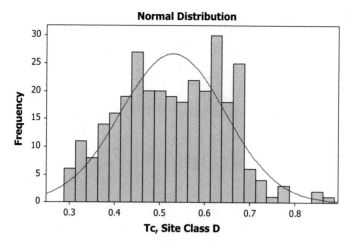

Figure A.3 Data frequency and normal distribution curves for site class D

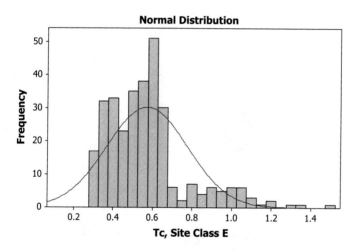

Figure A.4 Data frequency and normal distribution curves for site class E

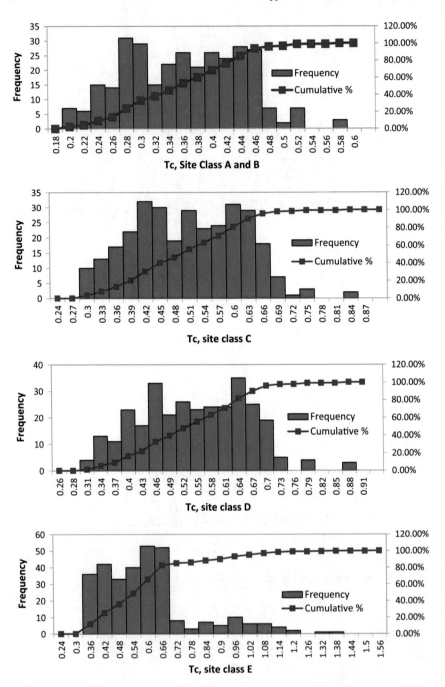

Figure A.5 Data frequency and cumulative curve

Reference

H-18-8 (2013) VA seismic design requirement. U.S. department of Veterans Affairs, Office of construction and facilities management

Appendix B
Masonry Material Parameters

This appendix provides details of the masonry material parameters that were used in the Finite Element Models presented in Chap. 5. The damage function considered in Chap. 5 for the masonry material model is presented in Fig. B.1.

Scaling of a failure surface

According to Malvar et al. (1997) if a new concrete with known unconfined compression strength, $f'_{c,New}$ is to be modeled, the a_i, k_i and c_i values, can be calculated as,

$$a_{0i} = a_0 r \tag{B.1}$$

$$a_{1i} = a_1 \tag{B.2}$$

$$a_{2i} = a_2/r \tag{B.3}$$

Moreover,

$$k_i = kr \tag{B.4}$$

$$c_i = cr \tag{B.5}$$

$$f_{ti} = f_t r^{2/3} \tag{B.6}$$

where

$$r = \frac{f'_{c,New}}{f'_{c,Old}} \tag{B.7}$$

© Springer International Publishing AG, part of Springer Nature 2019
R. Hassanli, *Behavior of Unbounded Post-tensioned Masonry Walls*,
Springer Theses, https://doi.org/10.1007/978-3-319-93788-5

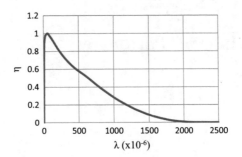

Figure B.1 Damage function obtained from masonry prism calibration

The input values in the format required for LS DYNA are presented in Table B.1. The values provided in Table B.1 were adopted for $f'_m = 13.3$ MPa. However, for a masonry material with f'_m other than 13.3 MPa, the material parameters were modified using Eqs. B.1–B.6. Note that instead of concrete, the compressive strength of masonry was considered in Eq. B.7. This was done for the five other values of f'_m that completed the set of walls tested by Laursen (2002) that were used to calibrate the FEM in Chap. 5. All other parameters were considered to be the same as for $f'_m = 13.3$ MPa, as presented in Table B.1.

Table B.1 Material parameters for $f'_m = 13.3$ MPa

$f'_m = 13.3$

$#	mid	ro	pr					
	2	2.00E−03	0.2					

$#	ft	a0	a1	a2	b1	omega	a1f	
	1.688	3.93	0.4463	6.08E−03	1.45	0.5	0.4417	

$#	slambda	nout	edrop	rsize	ucf	lcrate	locwidth	npts
	100	2	1	3.94E−02	145	0	25	13

$#	lambda1	lambda2	lambda3	lambda4	lambda5	lambda6	lambda7	lambda8
	0	8.00E−06	2.40E−05	4.00E−05	5.60E−05	7.20E−05	8.80E−05	6.00E−04

$#	lambda09	lambda10	lambda11	lambda12	lambda13	b3	a0y	a1y
	3.00E−03	1.00E−02	0.1	1	10	1.15	2.969	0.625

$#	eta1	eta2	eta3	eta4	eta5	eta6	eta7	eta8
	0	0.85	0.97	0.99	1	0.99	0.97	0.55

$#	eta09	eta10	eta11	eta12	eta13	b2	a2f	a2y
	5.00E−02	5.00E−02	5.00E−02	0	0	−5	4.90E−02	1.94E−02

*EOS_TABULATED_COMPACTION_TITLE

Grouted

$#	eosid	gama	e0	vo				
	1	0	0	1				

$#	ev1	ev2	ev3	ev4	ev5			
	0	−1.50E−03	−4.30E−03	−1.01E−02	−3.05E−02			

$#	ev6	ev7	ev8	ev9	ev10			
	−5.13E−02	−7.26E−02	−9.43E−02	−0.174	−0.208			

$#	c1	c2	c3	c4	c5			
	0	10.155	22.135	35.535	67.5			

(continued)

Table B.1 (continued)

$f'_m = 13.3$

$#	c6	c7	c8	c9	c10
	101.849	144.5	221.050	1290.5	1974
$#	t1	t2	t3	t4	t5
	0	0	0	0	0
$#	t6	t7	t8	t9	t10
	0	0	0	0	0
$#	k1	k2	k3	k4	k5
	6770	6770	6865	7210	8575
$#	k6	k7	k8	k9	k10
	9950	11315	12355	27795	33845

References

Malvar LJ, Crawford JE, Wesevich JW, Simons D (1997) A plasticity concrete material model for DYNA3D. Int J Impact Eng 19(9–10):847–873

Laursen PPT (2002) Seismic analysis and design of post-tensioned concrete masonry walls. Ph.D. thesis, University of Auckland, Department of Civil and Environmental Engineering, Auckland, New Zealand

Appendix C
More Details of the Experimental Study

This appendix provides details of the experimental study that was presented in Chap. 7.

- Construction and detailing of the precast concrete footing and loading beam (Figs. C.1, C.2, C.3, C.4 and C.5)
- Wall construction (Figs. C.6 and C.7)
- Preparation and testing of samples (Figs. C.8 and C.9)
- Wall assemblage and test monitoring (Figs. C.10 and C.11)
- Test instrumentations (Figs. C.12 and C.13)
- Photos of wall showing damage (Figs. C.14, C.15, C.16, C.17 and C.18)
- The history of measured gap opening and force in PT bars (Figs. C.19, C.20, C.21, C.22, C.23, C.24, C.25 and C.26)
- Calculation of energy dissipation (Fig. C.27)
- The accuracy of the proposed simplified and iterative method (Fig. C.28)
- Bilinear idealization of the capacity curves (Fig. C.29)

As shown in Fig. C.12, strain gauges were used to determine the strain history during testing. The forces in the PT bars of walls were measured during the tests using load cells located on top of the concrete bond beam through which the PT bars were past (Fig. C.12). As shown in the figure, the PT bars were secured to the bond beam. To be able to pass the PT bar strain gauge wires, the PT plates were recessed.

Equivalent viscous damping and energy dissipation

As described in Chap. 7, the equivalent viscous damping ratio, ξ_{eq}, can be calculated as

$$\xi_{eq} = E_d/(4\pi E_s) \tag{C.1}$$

where E_s is the stored strain energy and E_d is the dissipated energy as shown in Fig. C.27a.

© Springer International Publishing AG, part of Springer Nature 2019
R. Hassanli, *Behavior of Unbounded Post-tensioned Masonry Walls*,
Springer Theses, https://doi.org/10.1007/978-3-319-93788-5

Fig. C.1 Construction of precast concrete footing and loading beam

E_d was calculated as the area enclosed by a full cycle of response at each drift level (the first cycle was used in this study). The energy was calculated and plotted versus drift in Fig. C.27b. As shown in the figure, by increasing the applied drift, the dissipated energy increased in all wall tests.

Strength prediction of PT-MWs

The strength prediction ignoring the elongation of PT bar allowed by the MSJC (2013); and the proposed iterative method and simplified method which were presented in Chap. 7 (Table 7.3) are compared with the test results in Fig. C.28.

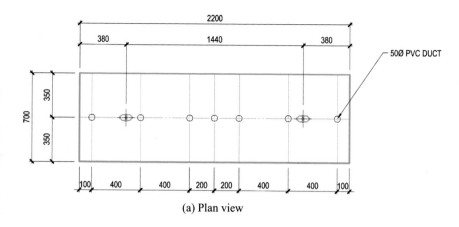

(a) Plan view

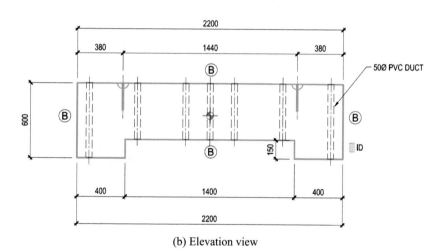

(b) Elevation view

Fig. C.2 Elevation and plan views of footing

The predicted and tested strength was denoted by $V_{predict}$ and V_{test}, respectively. As shown, while the MSJC (2013) is highly conservative in predicting the flexural strengths of the walls, both the iterative and simplified approaches provide a reasonable strength prediction. Note that unlike MSJC (2013), the elongation of PT bars are considered in the proposed iterative and simplified approaches.

Bilinear approximation curves

Bilinear approximation curves of the capacity curves that were summarized in Table 7.2 in Chap. 7 are presented in Fig. C.29.

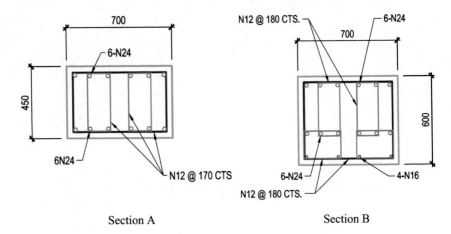

Section A   Section B

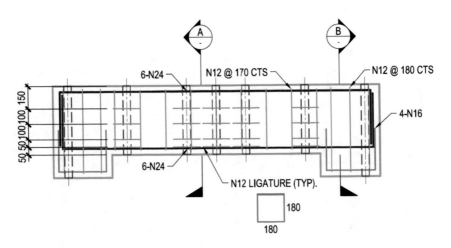

Fig. C.3 Footing reinforcement detail

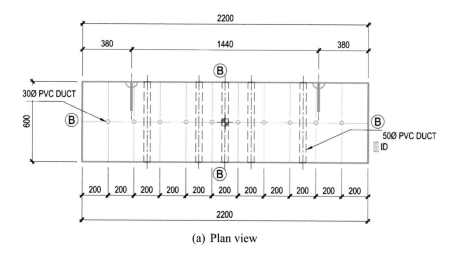

(a) Plan view

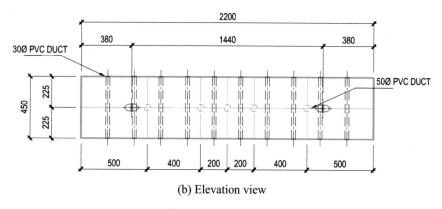

(b) Elevation view

Fig. C.4 Elevation and plan views of loading beam

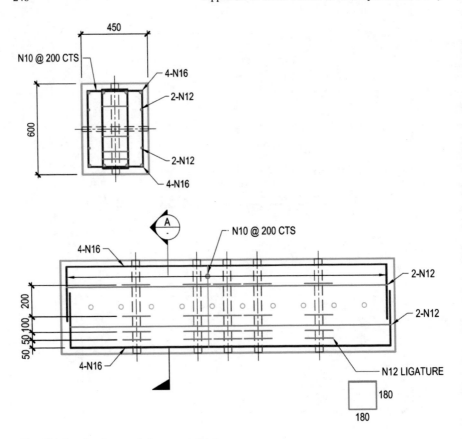

Fig. C.5 Loading beam reinforcement detail

Fig. C.6 Wall construction

Fig. C.7 Wall construction

Fig. C.8 Preparation of test material samples

Fig. C.9 Sample testing

Fig. C.10 Wall assemblage

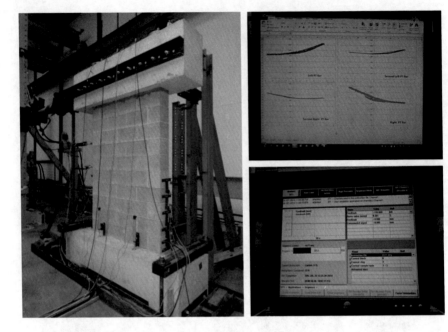

Fig. C.11 Test monitoring

Fig. C.12 Test instrumentation

Fig. C.13 LVDTs

(a)

(b)

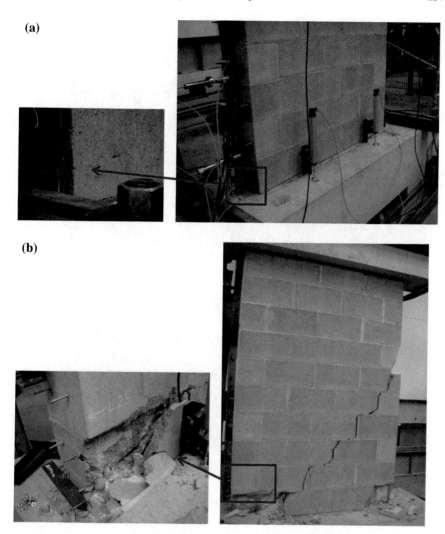

Fig. C.14 Damage of wall W1 **a** Flexural failure during testing, and **b** shear failure at the end of testing

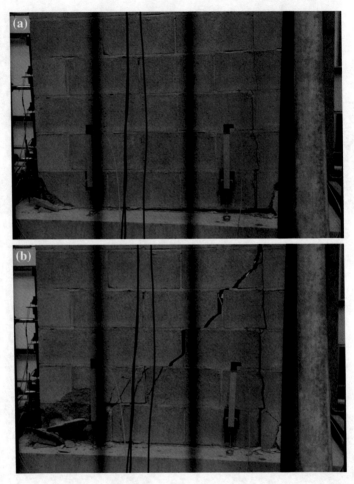

Fig. C.15 Damage of wall W2 **a** flexural failure during testing, and **b** shear failure at the end of testing

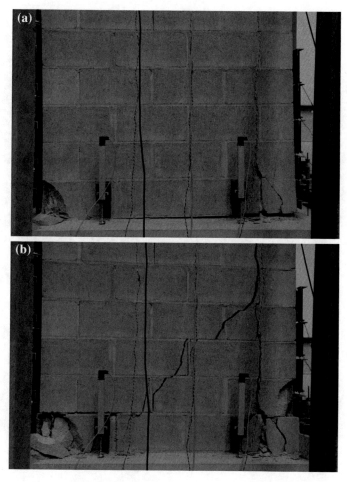

Fig. C.16 Damage of wall W3 **a** flexural failure during testing, and **b** shear failure at the end of testing

Fig. C.17 Damage of wall W4 **a** flexural failure during testing, and **b** flexural failure at the end of testing

Fig. C.18 Flexural failure of wall W4

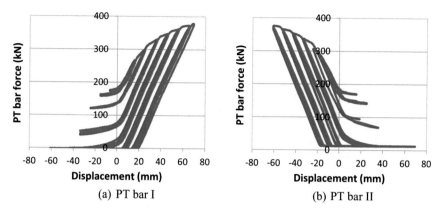

(a) PT bar I   (b) PT bar II

Fig. C.19 Force developed in PT bars of wall W1

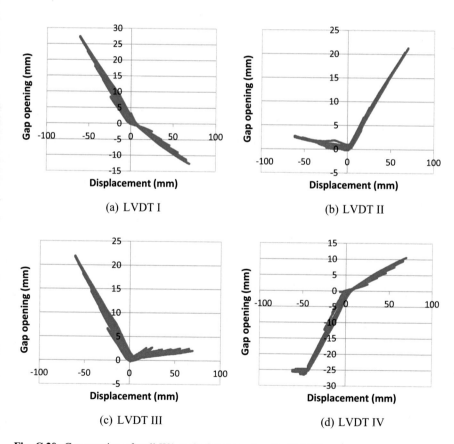

(a) LVDT I   (b) LVDT II

(c) LVDT III   (d) LVDT IV

Fig. C.20 Gap opening of wall W1 at the location of vertical LDVTs

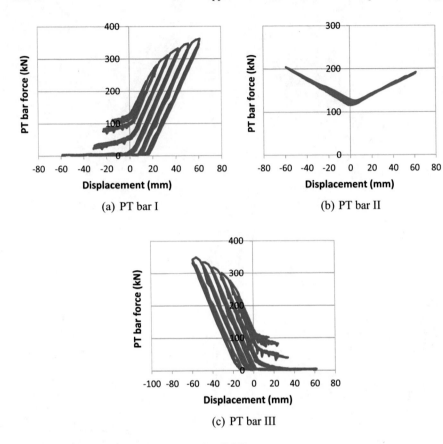

(a) PT bar I   (b) PT bar II

(c) PT bar III

Fig. C.21 Force developed in PT bars of wall W2

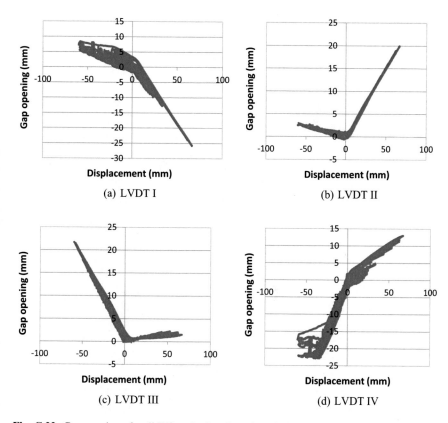

(a) LVDT I

(b) LVDT II

(c) LVDT III

(d) LVDT IV

Fig. C.22 Gap opening of wall W2 at the location of vertical LDVTs

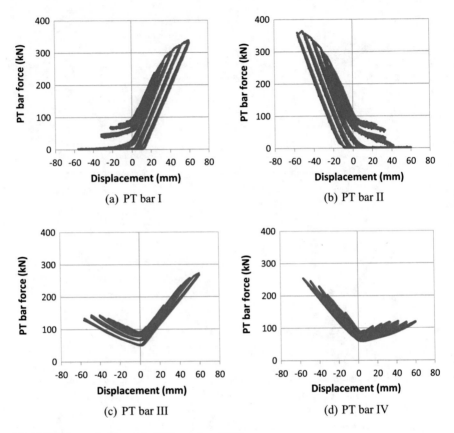

Fig. C.23 Force developed in PT bars of wall W3

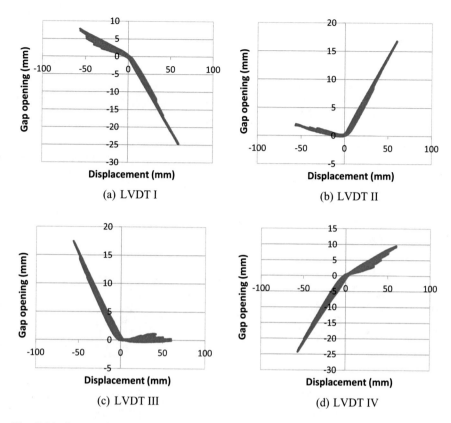

(a) LVDT I

(b) LVDT II

(c) LVDT III

(d) LVDT IV

Fig. C.24 Gap opening of wall W3 at the location of vertical LDVTs

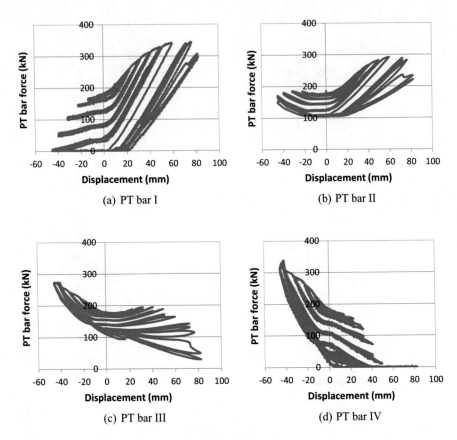

(a) PT bar I

(b) PT bar II

(c) PT bar III

(d) PT bar IV

Fig. C.25 Force developed in PT bars of wall W4

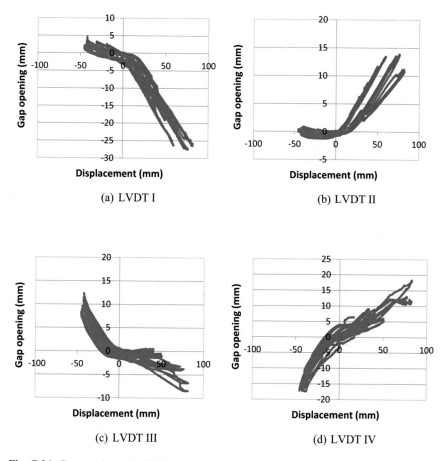

(a) LVDT I

(b) LVDT II

(c) LVDT III

(d) LVDT IV

Fig. C.26 Gap opening of wall W4 at the location of vertical LDVTs

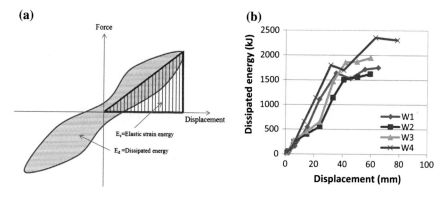

Fig. C.27 **a** Calculation of energy dissipation, and **b** energy dissipation in each cycle of tests

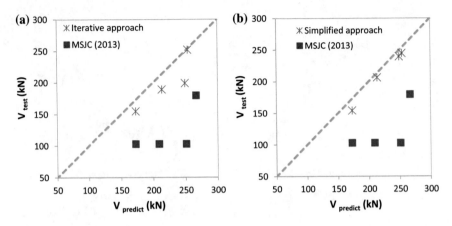

Fig. C.28 Predicted strength versus tested strength and MSJC (2013) method using **a** proposed iterative method and **b** proposed simplified method

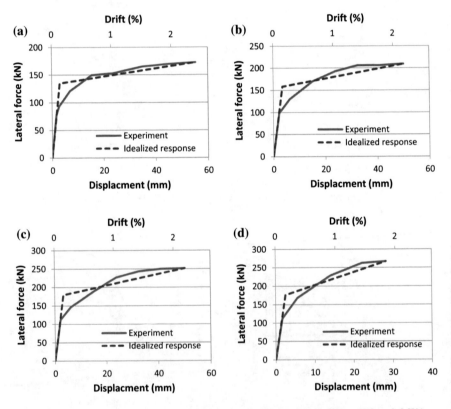

Fig. C.29 Bilinear idealization of the capacity curve of wall **a** W1, **b** W2, **c** W3 and **d** W4

About the Author

Dr. Reza Hassanli is currently a structural engineering lecturer at the University of South Australia. Reza has published number of journal articles and peer-reviewed conference papers, given invited seminars at several workshops and institutions. His research interests include: behaviour of precast concrete members, structural dynamics and seismic behaviour of composite and concrete structures, seismic rehabilitation of bridges and buildings, and post-tensioned systems and green concrete. Reza has received many grants and awards including UniSA Early Career Researcher International Travel Grant, Future Industries Accelerator (FIA) grant, NBERC Seed Funding, Mawson Lakes Fellowship Program (MLFP) and Australian Building Codes Board (ABCB) Scholarship.

© Springer International Publishing AG, part of Springer Nature 2019
R. Hassanli, *Behavior of Unbounded Post-tensioned Masonry Walls*,
Springer Theses, https://doi.org/10.1007/978-3-319-93788-5

Printed in the United States
By Bookmasters